Einstein's Masterwork

Einstein's Masterwork

1915 and the General Theory of Relativity

JOHN GRIBBIN

with MARY GRIBBIN

PEGASUS BOOKS

NEW YORK LONDON

EINSTEIN'S MASTERWORK

Pegasus Books Ltd
148 West 37th Street, 13th Floor
New York, NY 10018

Copyright © 2016 by John Gribbin

First Pegasus Books hardcover edition September 2016

ISBN: 978-1-68177-212-7

10 9 8 7 6 5 4 3 2 1

Printed in the United States of America
Distributed by W. W. Norton & Company, Inc.

Contents

List of Illustrations

The most valuable theory of my life ...
The theory is of incomparable beauty.

ALBERT EINSTEIN, 1915

If Einstein had not produced the Special
Theory in 1905, someone else would have
done so within a short time, five years or so.
That is the general situation with almost all
scientific advances, they are in the air; if A
doesn't make them, then B certainly will.

The General Theory is the startling exception,
maybe the only one in [the 20th] century. It is
agreed by the most eminent of theoretical physicists
– Dirac has said so without qualification – that if
Einstein had not created the General Theory [in
1915] no one else would have done so, perhaps
not until now, perhaps not for generations.

C.P. SNOW, IN *CONVERSATIONS WITH EINSTEIN*
BY ALEXANDER MOSZKOWSKI

About the Author

John Gribbin was born in 1946 in Maidstone, Kent. He studied physics at the University of Sussex and went on to complete an MSc in astronomy at the same university before moving to the Institute of Astronomy in Cambridge, to work for his PhD.

After working for the journals *Nature* and *New Scientist*, he has concentrated chiefly on writing books on everything from the Universe and the Multiverse to the history of science. His books have received science-writing awards in the UK and the US. His other biographical subjects include Erwin Schrödinger, Stephen Hawking, Richard Feynman, Galileo, Buddy Holly and James Lovelock.

Since 1993, Gribbin has been a Visiting Fellow in Astronomy at the University of Sussex.

Introduction
to the American Edition

In 1905, Albert Einstein published four scientific papers that had a profound influence on the science of the 20th century. Everybody knows Einstein's name, and an equation from one of those papers, $E = mc2$, is the most famous equation in all of science. For another of his contributions that year, he received the Nobel Prize. All of this has resulted in 1905 being referred to as Einstein's *"annus mirabilis,"* or "miraculous year." If he had never done another stroke of scientific work after 1905, Einstein would still be remembered as a genius. But, amazing as these achievements were, none of this represented Einstein's greatest work. Exactly ten years later, in 1915, he presented his master-work to the Prussian Academy of Sciences—a theory of gravity, matter, space and time which we know as the "General Theory of Relativity," and which he described as "the most valuable theory of my life." It describes the evolution of the Universe, black holes, the behavior of orbiting neutron stars, gravitational lensing, and why clocks run slower on the surface of the Earth than in space. It even suggests the possibility of time travel. He completed this work in Berlin during the First World War, where he later suffered from malnutrition caused by food shortages resulting from the Allied blockade of Germany and was nursed by his cousin, Elsa, who became his second wife. The accuracy of his theory was confirmed by British astronomers, at a time when Britain and Germany were technically still at war. But even today the General Theory is less feted than the achievements of 1905, because it is regarded as "too difficult," for ordinary mortals to comprehend. I hope to disabuse you of the year 1915 should be at least as celebrated as those of 1905.

The Special Theory of Relativity, one of the achievements of 1905, is "special" in the sense that it is restricted and "only" describes the behavior of things moving in straight lines at constant speed. The names alone tell you that the General Theory is a bigger deal, but because of the widespread (mis)conception that the General Theory is too difficult for ordinary mortals to understand, the events of 1915 have been less feted than the events of 1905.* It is, in fact, easy to understand the basics of the General Theory, even if the equations have been taken on trust, and this understanding should convince you—correcting the misconception that the Special Theory was Einstein's greatest achievement—that Einstein's greatest year was indeed 1915, not 1905. But I intend to demonstrate this by putting Einstein's science in the context of his life and work both before and after 1915, including his breakthrough year of 1905.

* * *

In February 2016, just as this edition of the book was going to press, a large team of scientists announced the first direct detection of the gravitational radiation predicted by Einstein in 1916, and described here in Chapter 4. This is seen by many as the ultimate proof of the accuracy of the General Theory of Relativity, although we already had firm evidence for the existence of such radiation from the behavior of objects known as binary pulsars. The experiment involves two laser beams, 4 km long, at right angles to each other, which are kept "in step" so that when their waves meet they cancel each other out and leave nothing. A pulse of gravitational radiation passing through the

---

* A point of pedantry: always the "General Theory," never "general relativity;" it is the theory that is general, not the relativity.

experiment gave a squeeze and stretch to the beams, so that they were briefly out of step, by a distance much less than the diameter of an atom, so that they interfered with one another and produced a measurable flicker. And an identical experiment many kilometers away saw an identical flicker just seven-thousandth of a second later, proving that the source was extraterrestrial, not caused by some lorry passing by on the highway, and travelled at the speed of light.

Analysis of the data showed that the pulse was produced by the collision of two black holes, each with the mass of about 30 Suns, in a galaxy more than a billion light years away. The collision released energy equivalent to the mass of three Suns, in line with the famous equation $E = mc^2$. The confirmation of Einstein's prediction was timely, but expected. What is even more exciting is that the ability to detect gravitational radiation from such events opens up a new window on the Universe. With more sensitive detectors already being planned and built, our understanding of the Universe is about to be transformed in the way it was transformed by the opening up of radio astronomy and X-ray astronomy. The legacy of Einstein's masterwork is very much part of 21st century science.

* * *

As ever, I am grateful to the University of Sussex and the Alfred C. Munger Foundation for providing, respectively, a base for me and a contribution towards my travel and other expenses. Thanks also to Estelle Asmodelle for help tracking down sources.

John Gribbin
http://johngribbinscience.wordpress.com

1   In the Beginning

Early life; Breaking free; Einstein and the Poly;
Rejection; Rescue

In 1905, Albert Einstein produced the most important
package of ideas from any scientist since Isaac Newton.
The iconic image we have of Einstein is the white-haired
genius, a wise and fatherly guru, a cross between God
and Harpo Marx. But in 1905 Einstein was a handsome,
dark-haired young man (he celebrated his 26th birthday
on 14 March that year), previously something of a ladies'
man but recently married, with a baby son. He didn't even
have a PhD at the time he published the scientific papers
that made him famous.

What made Newton and Einstein so special was that
they didn't just have one brilliant idea (like, say, Charles
Darwin with his theory of natural selection) but a whole
variety of brilliant ideas, within a few months of one
another. There were other similarities between the two
great men. In 1666, Newton celebrated his 24th birthday,
and although he had already obtained his degree from the
University of Cambridge, for much of 1665 and 1666 he
had been unable to take up a Fellowship at Trinity College
because the university had been closed by an outbreak of

plague. So he had been working in isolation at the family home in Lincolnshire. In 1905, the 26-year-old Einstein had already graduated from the Swiss Federal Polytechnic in Zurich, but had been unable to obtain a post at any university. So he had settled for a junior post at the patent office in Bern, working in isolation on scientific topics at home in his spare time – and also, as he later admitted, during office hours.

Especially in the theoretical sciences and mathematics, it is often true that people do their best work in their twenties, even if that work never matches the achievements of a Newton or an Einstein. But there the similarities between the two geniuses stop. Newton was a loner by choice, who made few friends and never married; although most of his great work was done in 1665–6, it was only published later, at different times, in response to pressure from colleagues who became aware of what he had achieved. Einstein was a gregarious family man, eager to get a foothold in the academic world, who knew the importance of advertising his discoveries and published them as soon as he could. It was his only chance of getting out of the patent office and into a university post. But what was the man later regarded as the greatest genius of the 20th century doing working in a patent office anyway?

Early life

Albert had been born in Ulm, in Germany, in 1879. In the summer of the following year, however, the family moved to Munich, in the south of Germany, where Albert's father, Hermann, went into partnership with his younger brother Jakob in the booming electrical industry. Jakob had a degree from the Stuttgart Polytechnic Institute and provided the expert know-how for the business; the money to set them up came from the family of Albert's mother, Pauline. Hermann's role was on the administrative side, running the business. Jakob and his wife, Ida, shared a pleasant house on the outskirts of Munich with Hermann, Pauline and little Albert.

Far from showing any signs of precocious genius, little Albert was so slow to learn to speak that his parents feared that there might be something wrong with him. It wasn't until well after his second birthday that he began talking, but when he did he used proper sentences from the start, quietly working the words out in his head and whispering them to himself before speaking out loud. In November 1881, his sister Maja was born, and she later recalled his response to her arrival, reported to her by her mother when she was old enough to understand. It seems that Albert had been told he would soon have something new to play with, and was expecting a toy; on being introduced to his baby sister, he asked: 'Where are the wheels?'* Although brother and sister developed an affectionate bond, young Albert was prone to occasional

---

* Maja's reminiscences are preserved in the Einstein archive at Princeton University.

outbursts of violent temper, when he would throw things at the nearest person – all too often, Maja. Family legend tells of the time he hit her over the head with a garden hoe, and how at the age of five he chased his first violin teacher out of the house, throwing a chair after her. But the music-loving Pauline was tough and strong-minded enough to ensure that he carried on with the violin lessons whether he liked them or not, eventually instilling in him a love of music that provided a lifelong release from the strains of his scientific work. And he had talent – much later, when he was a sixteen year old at high school in Switzerland, a school inspector would single him out for praise, reporting that 'one student, [named] Einstein, actually sparkled [in] his emotional performance of an adagio from a Beethoven sonata.'[1]

The expression 'tough love' could have been invented to describe Pauline's attitude to her children. When he was only four, Albert was given a guided tour of the neighbourhood by his parents, and from then on was not only allowed but encouraged to go out alone and find his way through the streets – although, unknown to the boy, they kept a distant eye on him on his first few solo expeditions. One of the first regular journeys he had to make was to school. The Einsteins were secular Jews and were unconcerned that the nearest school was Catholic, so that is where Albert received his first formal education, very much in the old-fashioned tradition of learning by rote and with strict discipline enforced by corporal punishment. Albert's disenchantment with the school was strengthened when he was still only five years old and ill in bed. His father gave the boy a magnetic compass to relieve his boredom, and Albert became intrigued by the way in which the needle always tried to point to the north, no matter how he twisted and turned the instrument. He was fascinated by the idea of an invisible force that kept a grip on the compass needle, and baffled that none of his teachers at the school had shown him anything half as interesting. This helped to instil an early conviction that he was much better off working things out for himself than working within the system.

Albert's stubborn insistence on finding his own way in the world led to a curious development during his years at the Catholic school. At that time, the statutes of the city of Munich required all students to receive some religious education, and although Hermann and Pauline didn't mind Albert attending a Catholic school, they drew the line at having him indoctrinated in the Catholic faith. So, to meet their obligations they got a relative to teach Albert about the Jewish faith – as they thought, just as a matter of form. To everyone's surprise, Albert lapped it all up and became something of a religious fanatic, observing the Jewish rituals that his parents had abandoned, refusing to eat pork and making up hymns that he would sing to himself as he walked to school in the morning. This religious phase lasted until Albert was about twelve and had been a student at the local high school (*Gymnasium*) for two years. His loss of faith was a direct result of his discovery of science; but that discovery owed nothing to the high school, and everything to a young medical student called Max Talmey.

Although Hermann and Pauline Einstein did not follow all the religious traditions of their nominal faith, there was one Jewish custom that they kept. At that time, there was a tradition among middle-class Jewish families of helping young students who might be struggling to make ends meet, and the Einsteins got into the habit of inviting Talmey, who came from Poland,* to dinner once a week. It was Talmey who introduced Albert, who was ten when they met, to the latest scientific ideas, discussing them with the boy as if he were an adult. Talmey lent books popularizing science to Albert, introduced him to algebra and in 1891 gave him a book about geometry. Einstein later described reading this book as the single most important factor in making him a scientist. He was gripped, fascinated by the way in which mathematical logic could be used to start from simple premises to construct truths, such as Pythagoras' Theorem, that are absolutely true. Within a year, he had worked his own way through the entire mathematics syllabus of the high school. This rather left the school, as he saw it, as a pointless waste of time. He had also lost his faith in God and now, as a young teenager, saw religion as part of a

---

* His name is sometimes translated as 'Talmud'.

deception played by the State in order to manipulate its people, especially the young.

Albert had made a good start to his time at the Luitpold Gymnasium, finding it easy to keep up with his peers; but although this was one of the best schools of its kind, a combination of the rigid educational system, his loss of religious faith and his discovery that he could leap ahead of the curriculum by working on his own led him to neglect his classwork. He still got good grades in mathematics, but couldn't see the point of subjects like Classical Greek and gained a reputation as an impudent troublemaker. One family anecdote recalls that Albert's father was summoned to the school to be told that his son was a troublemaker. When he asked what they meant, he was told 'he sits at the back and smiles'.

The situation at school came to a head in 1894, when the family business went bust. The firm had done reasonably well in the 1880s, like many other small businesses taking advantage of the booming demand for electrical equipment such as dynamos, lighting systems and telephones. But in the classic pattern of progress following the invention of a new technology, those small businesses were now being swallowed up by large firms, or going under in the face of competition from the giants, such as Siemens and AEG. It was a lack of capital, as much as anything else, that saw the Einsteins lose out in this competition. But before matters came to a head they had gained a reputation in southern Germany and northern

Italy. Italy had not gone so far down the road towards domination of the electrical industry by one or two firms, and with the encouragement of one of their Italian colleagues Hermann, Jakob and their wives decided to move to Italy to start again. There was one snag. Albert, who was now fifteen, had three years left to complete at the *Gymnasium*, which would ensure his admission to a good university. A family decision was made to leave him behind in Munich, staying in a boarding house but with a distant relative keeping an eye on him.

The result ought to have been predictable. Albert was lonely and miserable in Munich, and without family life to fall back on the school seemed unbearable. There was another worry. Albert had always hated the militaristic aspect of German society at the time, and as a young boy, scared by the sight and sound of marching troops, had once begged his parents to promise that he would never have to be a soldier. But the law then required every German male to undergo a period of military duty. The only escape was to leave the country and renounce his German citizenship before his seventeenth birthday; if he left later, he would be regarded as a deserter.

Breaking free

There is some confusion about how exactly Einstein engineered his escape from Munich, but this is the most likely pattern of events. He was certainly depressed and managed to persuade his doctor to provide a certificate stating that he should rejoin the family for health reasons. He also persuaded his maths teacher to provide a letter stating that Einstein had already mastered the syllabus, and there was nothing more he could teach the young man. Armed with these documents, Einstein approached the principal of the school and told him that he was leaving; the principal's response, we are told, was that Einstein was being expelled anyway, for being a disruptive influence. The likelihood is that Einstein had carefully cultivated his role as a disruptive influence in order to ensure that he would not be asked – let alone ordered – to stay on. Whether he jumped or was pushed, there is no doubt that in the spring of 1895, just six months after being left in Munich to complete his education, Einstein turned up on the doorstep of his parents' new home in Milan, in northern Italy.* He duly renounced his German citizenship (the declaration took effect in January 1896) and swore he would never go back to that country. By then, Einstein was living in Switzerland and set in motion the slow process of obtaining Swiss citizenship; but in those relatively free and easy days in Europe the lack of a passport did not prevent him travelling as he pleased, in particular between Switzerland and Italy. On official forms, under 'nationality' he simply described himself as 'the son of German parents'.

---

* Shortly after Albert arrived, the family moved to the smaller town of Pavia, near Milan.

In the summer of 1895 Einstein was sixteen, in Italy, with no responsibilities (and no prospects). Although he did some work for the family business and had vague notions of becoming a teacher of philosophy, for several months he mostly just enjoyed himself, touring the art centres of Italy, visiting the Alps and falling in love with the culture. When pressed by his father to settle down and give some thought to the future, Einstein assured Hermann that in the autumn he would take the entrance examination for the Swiss Federal Polytechnic in Zurich (often known by the initials of its German name, as the ETH). The ETH was not a great university in the mould of Heidelberg or Berlin, but a new kind of institution devoted primarily to the education of would-be teachers and engineers. The cocksure Einstein was certain he would be able to walk into an establishment with such relatively modest academic standards, and it came as a rude shock when he failed the exam. (His later claim to have failed deliberately, in order to avoid being pushed into a profession by his father, should be taken with a large pinch of salt; if that were the case, why would he have gone back and embarked on the same course a year later?)

In fact, Einstein was lucky to be allowed to take the exam. The normal age for this was eighteen, and applicants were expected to have a high school diploma; Einstein was still six months short of his seventeenth birthday and had left high school under a cloud. But the director of the ETH, Albin Herzog, recognised Einstein's potential and offered him a lifeline. He suggested that Einstein should enrol in a Swiss secondary school in the town of Aarau, a little way outside of Zurich, to do some catching up before taking the entrance exam again in 1896. There, he could lodge in the home of Jost Winteler, a teacher at the school, and live as one of the family.

The domestic arrangements suited Einstein down to the ground. As Winteler taught Latin and Greek he was not one of Einstein's tutors, so school did not intrude too much on daily life. One of Einstein's cousins, Robert Koch, was a student at the school and lodging next door. And there was plenty of family life – the Wintelers had three daughters and four sons, plus a couple of other paying guests, and Einstein was soon one of the family. A classmate during this year in Aarau later described the Einstein of 1895–6, to his biographer Carl Seelig, as 'sure of himself, his grey felt hat pushed back on his thick, black hair [striding] energetically up and down ... unhampered by convention, his attitude towards the world was that of the laughing philosopher, and his witty mockery pitilessly lashed any conceit or pose.'

Mature beyond his years, Einstein clearly made a big impact on his companions – none more so than Marie Winteler, the eighteen-year-old daughter of the house, who had just completed her course as a trainee teacher and was living with her parents while waiting to start her first job. In spite of the age difference, an enormous gulf for most teenagers, the two fell in love. Both families seem to have been happy with the development, which blossomed into something like an unofficial engagement and persisted after Einstein returned to Pavia and then moved on to Zurich. But in the spring of 1897, as his own horizons broadened and he made new friends in the city, he broke off the relationship. It took some time to convince Marie that he meant it, but in the end everything was settled amicably, and the Einsteins and the Wintelers remained good friends – so much so that Einstein's sister Maja later married Marie's brother Paul. Anna, another of the Winteler siblings, married Michele Besso, Einstein's best friend in Zurich and for long after his student days were over. In a letter written to Besso's wife and son after he died in 1955, Einstein said 'what I admired most about Michele was the fact that he was able to live so many years with one woman, not only in peace but also in constant unity, something I have lamentably failed at twice.'[2]

Alongside his happy relationship with the Winteler family, an active social life, music and his first experience of love, Einstein did enough work at school (where he had been allowed to join the final year group, with classmates a year older than himself) to achieve high grades in all his subjects except French, which he always struggled with. In the Swiss system, examination papers were marked on a grading scale from 1 to 6, with 6 being the top mark. Einstein's average of 5½ was the best in his year, from the youngest pupil in the class. Although this was brought down by his French paper, which was rather generously graded between 3–4, the essay he wrote (in French) for the examination is the most interesting of the papers (which survive and can be found in *The Collected Papers of Albert Einstein*) because the set subject of the essay was '*Mes Projets d'Avenir*'.* Ignoring the terrible French, the essay gives us a glimpse into Einstein's personal ambitions at the time, which seem remarkably limited in the light of what he would achieve:

> If I am lucky enough to pass my examinations, I will attend the Polytechnic in Zurich. I will stay there four years to study mathematics and physics. My idea is to become a teacher in these fields of natural science and I will choose the theoretical part of these sciences.

Einstein goes on to say that this ambition is partly based on the fact that he lacks any 'practical talent', but also because 'there is a certain independence in the profession of science that greatly appeals to me'. That certainly chimes with the way his life would develop. But there was already a hint of what was to come in some of the ideas Einstein did not commit to paper in 1896. He later recalled that while still a schoolboy in Aarau, he puzzled over the idea that if you could run at the speed of light, you would see a light wave standing still alongside you, frozen in

---

* My Plans for the Future.

time, as it were; but the laws of physics said that such a 'time-independent wave-field' could not exist.[3] It would be nine years before he found the solution to that puzzle.

Einstein and the Poly

After passing his entrance examination for the ETH early in the summer of 1896, Einstein spent some time with his family in Italy, where the electrical business was once again in crisis. Jakob left to work for another company and ended up living comfortably in Vienna as the manager of a firm of instrument makers; Hermann tried to make another fresh start in Milan. Albert was sufficiently concerned about the prospects for this latest venture that he tried unsuccessfully to dissuade his relations from pouring more money into the scheme; but in October he had to put these family difficulties behind him as he returned to Zurich to enrol for his course.

In spite of the year spent in Aarau, Einstein was still six months short of the official age for admission, eighteen, and one of the youngest students ever admitted to the ETH. The 'Poly', as it was known locally, wasn't that old itself, having been founded in 1855 as the first university-level academic institution in Switzerland (the Swiss Confederation only came into being in 1848). Since then, three universities had been established in Switzerland – in Basel, Zurich and Geneva. Unlike the Poly, which was a Federal Swiss government establishment, the universities were run by their respective cantons. Unlike the universities, initially the Poly could not award doctoral degrees; but in 1911 it was given full university status including the right to award doctorates. That hasn't stopped it being known as the Poly right down to the present day.

At the end of the 19th century, there were just under a thousand students at the ETH. But, as its name implies, the ETH was primarily devoted to the education of engineers, not theoretical physicists, and there were just five students, including Einstein, taking the science course in his year. These included Marcel Grossmann, a model student who attended all the lectures and took detailed notes which he kept carefully for revision. Grossmann became a firm friend of Einstein, and in the long run those beautifully written notes would prove even more important to him than to Grossmann. The group also included a lone woman, Mileva Maric, a Serbian, from what was then part of the Hungarian Empire. Mileva had had to struggle with an unsympathetic family and unsupportive school system at home to make it to university in Switzerland, the only German-speaking country where women were admitted to university at that time. Indeed, Mileva was only the fifth woman to be admitted to study physics at the ETH. The other two members of the class were Jakob Ehrat, a hardworking but unspectacular student, and Louis Kollros; both, like Grossmann, were Swiss.

Like many students, Einstein enjoyed the freedom of university life to the full and didn't worry too much about the academic side until the examinations loomed. He didn't reckon anyone could teach him mathematics better than he could learn on his own with the aid of books, and he seldom attended the lectures, leading one of his professors, Hermann Minkowski, later famously to describe the student Einstein as a 'lazy dog', who 'never bothered about mathematics at all'. In fact, Minkowski was one of the few professors at the ETH that Einstein respected, and when he did attend a lecture given by Minkowski during his final semester at the ETH, Einstein remarked to Louis Kollros that 'this is the first lecture on mathematical physics we have heard at the Poly'.[4]

Cutting lectures gave Einstein plenty of time to indulge his passions for: coffee-house discussions with his friends setting the world to rights (including scientific discussions about the latest ideas in physics); the company of women (he always got on well with women, who were charmed by his manners, his music and his masculine good looks); sailing on the lake (where he always took a notebook in case the wind dropped, so that he could scribble down his ideas on physics); and music (combining this with his love of the company of women, Einstein often gave violin recitals in the homes of ladies of his acquaintance). He lived in lodgings in Zurich, getting by financially on an allowance of 100 Swiss francs a month, generously provided by one of his maternal aunts, and supplemented by a little private tuition. Out of this, he set aside 20 francs to save towards the fee he would have to pay when he was eventually awarded Swiss citizenship. On Sundays, he took lunch with the family of Michael Fleischmann, a Zurich businessman, echoing the way the Einsteins had looked after Max Talmey in Munich.

It was music that brought Einstein and his lifelong friend Michele Besso together. Besso was six years older than Einstein and already working as a mechanical engineer. They met at a house where Einstein was among the musicians entertaining a group of students and other people, an important, if informal, social activity in those days before TV, radio or recorded music. It was Besso who introduced Einstein to the work of Ernst Mach, an Austrian philosopher-physicist who had made important contributions to the scientific debate raging at the end of the 19th century concerning the reality of atoms.

From our modern perspective, it is hard to believe that only a little over a hundred years ago people were still arguing about whether or not atoms were real. But this was indeed an important debate, which would influence a great deal of Einstein's early scientific work and become a significant feature of his *annus mirabilis*.

Popular accounts of the history of science often tell you that the idea of atoms goes back to the time of the Ancient Greeks, but this is only true up to a point. What is true is that the Greek philosopher Democritus, who lived in the 5th century BC, did discuss the idea that everything is made of tiny, indestructible particles moving through a void (the vacuum) and interacting with one another.

But this was never more than a minority view at the time and was dismissed by most of the Ancient Greek thinkers because they could not accept the idea of the void, a genuine nothingness between atoms. The idea was revived from time to time, notably by the Frenchman Pierre Gassendi in the 17th century, but was always dismissed, for the same reason. It was only in the 19th century that a large group of scientists really began to take the idea of atoms seriously, and even then others argued against the idea.

The scientists who took the idea of atoms seriously found evidence supporting it in both chemistry and physics. In the early 1800s, John Dalton, in England, developed the idea that each element (such as hydrogen or oxygen) is made up of a different kind of atom (but with all the atoms of a particular element identical to one another), and that compound substances (such as water) are made up of molecules in which different kinds of atom join together (in this case, as H_2O). As early as 1811, jumping off from these chemical discoveries, the Italian Amadeo Avogadro announced his famous hypothesis, that at a given temperature and pressure, equal volumes of gas contain the same number of particles (molecules or atoms), with the clear implication that there is nothing in the space between these particles. But his idea was ignored for decades, and there was no clear idea of the differences between atoms and molecules until the work of Avogadro's compatriot, Stanislao Cannizaro, in the 1850s.

By then, evidence supporting the idea of molecules was coming in from the physicists. One of the most important practical applications of science in the 19th century concerned the study of heat and motion (known as thermodynamics), which was directly relevant to the application of steam power during the industrial revolution. By studying the way in which heat could be generated and how it flowed from one object to another, scientists came up with laws of thermodynamics to describe the relationship between work and energy on the scale of the kind of machinery that powered the Industrial Revolution – sometimes referred to as the 'macroscopic' scale. Physicists such as the Scot James Clerk Maxwell, Hermann von Helmholtz in Germany and the Austrian Ludwig Boltzmann then developed models to describe these macroscopic phenomena in terms of the accumulated effect of huge numbers of atoms and molecules bouncing around and interacting with one another like tiny, hard spheres, obeying the basic laws of mechanics discovered by Isaac Newton 200 years earlier. This behaviour of atoms and molecules at a lower level is often referred to as 'microscopic' behaviour, but, crucially, atoms are actually far too small to be seen by any microscope available in the 19th century.

The way the cumulative behaviour of vast numbers of atoms and molecules interacting on the microscopic scale combines to produce measurable effects on the macroscopic scale is called 'statistical mechanics'. For the particular case of gas trapped in a box, this approach proved an excellent way to explain how the pressure and temperature of the gas change as the box is made smaller or larger and the speed of the molecules and atoms changes; this is known as the 'kinetic theory', since it is all about movement.

All of these ideas were in the air in the 1890s and formed the subject of many conversations between Einstein, Grossmann, Besso and their friends, wreathed in tobacco smoke as they lingered over coffee in some Zurich café, or while striding through the countryside on extended walks. The problem was that although ideas like statistical mechanics and the kinetic theory worked at a practical level to provide a mathematical description of what was going on, nobody had seen atoms – more to the point, given the technology of the time it was physically impossible to see atoms. This left the door open for philosophers such as Mach to argue that the atomic hypothesis was no more than a hypothesis, what is known as a 'heuristic device', meaning that just because things in the macroscopic world behave as if they were made of atoms that doesn't prove that they are made of atoms. Mach regarded atoms as no more than a convenient fiction, which provided a basis for physicists to make calculations; anything that could not be detected by the human senses, he argued, was not the proper subject of scientific debate.

Einstein disagreed and argued the case for atoms with his friends. He became obsessed with the idea and determined that if no one else could prove that atoms were real, he would do it himself. As he wrote many years later in his *Autobiographical Notes*, he determined that as soon as he had graduated from the ETH he would search for evidence 'which would guarantee as much as possible the existence of atoms of definite finite size'. He succeeded, as we shall see in the next chapter. Important as that success was, though, just as important as the fact that he succeeded is the fact that at the beginning of the 20th century Albert Einstein, widely regarded as the greatest genius of the 20th century, thought that the most important problem facing science was to prove the reality of atoms. That alone shows just how far we have come in the past hundred years.

At first, Einstein did well at the ETH, his brilliance enabling him to shine in his intermediate examinations even though he had not been following the curriculum as diligently as his friend Grossmann. But alongside his studies and his discussions about the latest hot topics in science he developed another passion, one which meshed with his interest in physics but would eventually prove a distraction from his scientific work. It isn't clear exactly how or when Einstein and Mileva Maric became more than just friends, but their discussions about physics seem to have developed a romantic side by the early summer of 1897. It may be no coincidence that Einstein broke off his relationship with Marie Winteler in the spring of that year, but Mileva seems to have been the first to be seriously affected by the new relationship.

The circumstantial evidence suggests that at about this time Mileva began to fall in love with Einstein and became confused about her future prospects. She had made a huge effort, as a woman, to get a place at the ETH – a foothold on a professional career in science – and seems to have become concerned at the possibility that it might all be wasted if she were to marry, settle down and (as would have been almost inevitable in those days) have children. In the summer of 1897, she went home to her family as normal and kept up a correspondence with Einstein; but at the end of the summer, instead of returning to Zurich she went, without explanation, to Heidelberg. Although this was the home of one of the great German universities, it wasn't so good for women, who were patronisingly allowed to attend lectures by special permission of individual professors but were not allowed to take a degree. Whatever her motives for this diversion, by April 1898 Mileva was back in Zurich, where Einstein promised to help her catch up with her coursework (using Grossmann's notes); but, as we shall see, academically she never did make up for this lost time.

Einstein himself sailed through the 'intermediate diploma examination', held in October 1898, coming top of his small class with a grade average of 5.7; Grossmann came second with 5.6, and Mileva had to postpone the exam because of the time she had spent in Heidelberg. Einstein's promised help with her catching up proved to be more of a hindrance than a help, in more ways than one.

In his third year at the ETH, Einstein developed his interest in electromagnetism, and in particular the behaviour of light. Mileva was roped in and used as a sounding board for his ideas – much more exciting than the coursework – when she should have been concentrating on the curriculum. Maxwell (who, as it happens, died in the year Einstein was born, 1879) had discovered a set of equations which describe, among other things, the way electromagnetic waves move. The equations predicted that the speed of those waves would be 300,000 kilometres per second.* Since the speed of light was being measured accurately at around the same time that Maxwell came up with this number (in the 1860s), and the speed of light exactly matches the predicted speed of electromagnetic waves, this was seen as proof that light is a form of electromagnetic vibration.

But what was vibrating? By analogy with the way sound waves propagate as vibrations in the air, or in other substances, in the 19th century physicists thought that light (and other forms of electromagnetic wave) must propagate in the form of vibrations in a tenuous substance they called 'the ether'. The ether was assumed to fill all of space, and even somehow to fill the atmosphere of the Earth, mingling with the air, to enable light to propagate. It would, though, have to be very tenuous,

---

* More precisely, 299,792.458 km/second. In a logical world, the second would be redefined to make the speed of light exactly 300,000 km/sec.

since planets, and even people, seemed to be able to move through it as if it did not exist. And yet, since the speed with which waves travel depends on the stiffness of the substance they propagate in (so that sound travels faster in steel than in air), it would have to be incredibly stiff – far stiffer than steel.

By the end of the 19th century, unsuccessful attempts were being made to measure the speed of the Earth through the ether, by measuring the speed of light in directions along the line of the Earth's motion through space and at right angles to its motion; but the speed of light always seemed to be the same, whichever way the experimental apparatus was pointed. Einstein seems to have either been unaware of these experiments or unimpressed by them, but he was fascinated by Maxwell's equations and their prediction of a specific speed for light. He now had another 'thought experiment' to highlight the mysterious nature of these waves. If you could run at the speed of light while holding a mirror in front of your face, would you be able to see your reflection? Presumably not, since the light leaving your face to bounce off the mirror and come back into your eyes as a reflection would never be able to catch up with the mirror!

But there was something else about Maxwell's achievement that appealed to Einstein. Maxwell had worked his ideas out entirely theoretically. He had produced his equations and predicted the speed of light without doing an experiment. He had also predicted that there must be other forms of electromagnetic waves, what we now call 'radio waves', and these had duly been discovered by the German Heinrich Hertz in 1888. Or rather, Hertz had not discovered radio waves; he had merely detected what Maxwell, the theorist, had 'discovered'. Instead of experimenters making observations which the theorists then had to try to explain, the theorist had worked it all out on his own. This was what appealed to Einstein: the idea that the power of the human mind and mathematics was alone enough to conjure up deep truths about the world, echoing the way the Greeks had conjured up deep truths about geometry.

As early as the summer of 1899, Einstein wrote to Mileva that:

> I'm more and more convinced that the electro-dynamics of moving bodies as it is presented today doesn't correspond to reality, and that it will be possible to present it in a simpler way.[5]

And he said in the same letter that it was not possible 'to ascribe physical meaning' to the concept of the ether. But he seems to have been unable to take his ideas further at that time, and the discussion of electrodynamics lapsed during his fourth and final year at the ETH. Getting ahead of our story only slightly, the idea of the ether was indeed soon made redundant, and physicists now think of light and other forms of radiation in terms of 'fields', rather like the pattern of lines of force that you can see when a bar magnet is placed under a sheet of paper and iron filings are gently sprinkled on top of the paper. Once they are given a push, waves propagate through these fields until they interact with something else.

Einstein's own interaction with Mileva clearly intensified during their fourth year at the ETH, and it was probably sometime during that year that they decided to get married. They had spent the summer of 1899 apart, with Einstein on holiday with his mother and sister, and Mileva at home revising for her postponed intermediate examination. But although they still had separate lodgings in Zurich on their return, Einstein seems to have spent as much time at hers as in his own, and in a letter written in October 1899 refers to his 'household' with Mileva.[6] Soon, he was addressing her as 'Doxerl' ('Dolly', as in 'little doll') and she was calling him 'Johonzel' ('Johnny'). In spite of the deepening of their relationship and his neglect of the formal lectures throughout his time at the ETH, by cramming hard using Grossmann's notes Einstein was able to pass his final examinations. Not gloriously. He came fourth out of the five students sitting the exam, with a grade average of 4.9. But Mileva, who lacked his genius, was unable to make up adequately for lost time; she scored only 4.0 and failed. She did spectacularly badly at maths, in particular, scoring only 2.5 out of the possible 6 marks.

By now of course, Einstein's parents knew all (or perhaps, not quite all) about his relationship with Mileva.

In spite of his rebellious nature and lack of obedience to authority, their Albert had made it through the educational system to earn a diploma which ensured him a respectable job. But when he told them, in the summer of 1900, that he intended to marry 'Dolly', they were aghast. In their view he was far too young (only just 21), they weren't sure about the girl (who was in any case four years older than him), and they believed that a man ought to achieve security (not least, a job) before he even contemplated marriage. Even Einstein's friends in Zurich seem to have been surprised that a man who was so attractive to women should have chosen this one for his permanent partner. But part of the attraction to him seems to have been the thought of sharing his life with someone he could share everything with, including science. He fantasised in his letters to Mileva about them both getting PhDs and working together on scientific papers that would take the world by storm. But first, she had to retake her examinations at the ETH, and he had to get a job in Zurich at the Poly itself or at the University, so that they could be together.

Rejection

It was at this point that reality began to intrude on the dream. Einstein's diploma from the ETH qualified him to work as a teacher of mathematics and science in secondary schools. What he wanted was to get a job as an assistant to one of the professors and work for a PhD.

Although the ETH did not at that time award doctorates, there was an arrangement whereby any graduate from the ETH could write a dissertation for submission to the University of Zurich and be considered for the award of a PhD; so this was by no means an impossible dream, at least for the top students. But Einstein was not, on paper at least, one of the top students. He seems to have been convinced that in spite of obtaining the worst pass in his year and still having a reputation as a cocky, know-it-all troublemaker, his innate ability (of which he at least was in no doubt) would be recognised and be sufficient to obtain the post he wanted. But that was not the way the system worked. Hardly surprisingly, the men who had done best in the exams received priority in consideration for what limited opportunities were available, and in any case the authorities at the ETH were happy to see the back of Einstein, who had been far from a model student. They may well, also, have known about his relationship with Mileva, which went beyond the bounds of what was regarded as 'decent' in that time and place.

The couple again spent the summer apart, each with their respective families, and in the face of parental opposition talk of marriage ceased. Mileva studied for her resits, while Einstein applied unsuccessfully for assistant-ships, took a little holiday and tried to forget his troubles by studying Boltzmann's work on thermodynamics. In the autumn, both returned to Zurich where, in spite of the lack of any job for Einstein at the ETH, they both paid the appropriate fees and registered to work in the lab there on problems involving heat and electricity (the electron itself had only been discovered in 1897, after Einstein had begun his course at the ETH), with a view to obtaining a pair of PhDs. Einstein's allowance from his aunt had stopped when he graduated, and he had to make ends meet by private tutoring; Mileva still had a small allowance from her father. In these less than promising circumstances, Einstein completed his first scientific paper, on capillary action, which was duly published in the prestigious *Annalen der Physik*. The paper concerned the nature of the forces between molecules, and this led Einstein in a direction that would, eventually, result in the award of a doctorate. But Mileva's hopes of obtaining a PhD gradually faded away.

Apart from Einstein's first scientific paper, the other highlight of what must have been a far from happy few months in Zurich came in February 1901, when Einstein at last became a Swiss citizen. Somewhat ironically, in view of his reasons for renouncing German citizenship, he then had to take a medical examination prior to a spell of compulsory military service – but failed on the grounds of flat feet and varicose veins. So he probably would not have been forced to serve in the German army even if he had stayed in Munich.

Just a month after becoming a Swiss citizen, though, Einstein had to admit defeat in his efforts to find suitable employment in Zurich. Under pressure from his parents to get a real job (any job!), he abandoned the increasingly faint prospect of a dissertation on thermoelectricity and returned to Italy, leaving Mileva to work towards her second sitting of the final examinations (while also trying to keep her own studies of heat conduction going), and facing an uncertain future.

In Milan, Albert at least had Michele Besso, who now lived in that city with his wife and young daughter, to talk to about physics. He also came across the new work of Max Planck, a German physicist who had just discovered that some of the properties of light and other forms of radiation could best be explained if the radiation is only emitted or absorbed by objects in the form of discrete packets of energy, which became known as 'quanta'. His key paper, laying the foundations of what became quantum physics, had only been published in 1900, but as usual Einstein was keeping up with cutting-edge research in physics.

He was also keeping up his older interests, including his interest in light and motion. His letters to Mileva are full of the usual endearments expressed by separated lovers, but they also include references to his scientific work, and one of those references in particular has led some people to see Mileva as a more important scientific influence on him than she really was. In a letter written in April 1901 he writes: 'I'll be so happy and proud when we are together and can bring our work on relative motion to a successful conclusion.'[7] Taken out of context, you might see that as indicating that Mileva made a significant contribution to the theory of relativity. But the proper context is that she failed her exams (not least because she spent too much time discussing esoteric ideas with Einstein instead of concentrating on her proper coursework), and she was by far the weakest mathematician in her year. She was alone in Zurich, struggling (in the end, unsuccessfully) to prepare for her resits, and Einstein was trying to cheer her up by assuring her how important she was to him. In the same letter, he writes: 'you are and will remain a shrine for me to which no one has access; I also know that of all people, you love me the most and understand me the best.' He was clearly besotted with her and saw the work on what became relativity theory as 'ours' in the sense that 'what's mine is yours'. But there is not one shred of evidence that Mileva contributed anything more to the development of Einstein's great theory than her role as a scientifically literate listener on whom he could try out his ideas.

Soon after that letter was written, things began to look up for Einstein. The first good news came in a letter from his old friend Marcel Grossmann. Grossmann had mentioned Einstein's increasingly desperate search for a job to his father, who happened to be a friend of the director of the Swiss Patent Office, Friedrich Haller. Haller had told the elder Grossmann that there would shortly be a vacancy at the patent office in Bern, and that Einstein should certainly apply for the post when it was advertised.

Just when the post was likely to be advertised was not clear, but the very next day Einstein received another letter offering him some temporary work. Jakob Rebstein, who had formerly been an assistant at the ETH and knew both Einstein and Mileva, now had a job as a mathematics teacher in Winterthur, near Zurich, but was about to go on his spell of compulsory military service for a couple of months. Would Einstein like to act as his locum?

The job would not only help to fill the gap while waiting for the patent office post to be advertised, but meant that he would be near Mileva. To celebrate the good news, the couple met up at Lake Como for a short holiday before he took over from Rebstein. In a letter to a friend, Mileva described how they rode through the snow-covered countryside in a horse-drawn sleigh:

> We drove one moment through long galleries and the next on the open road, where, all the way to the remotest distance, our eyes could see nothing but snow and more snow, so that at times I shuddered at this cold white infinity and firmly kept my arm round my sweetheart under the coats and blankets which covered us … I was so happy to have my lover for myself again for a while, the more so as I saw that he was just as happy.[8]

Soon after this trip Mileva must have told Einstein that she was pregnant, since towards the end of May 1901 his letters to her start referring obliquely to the happy event, assuring her that he would stand by her and all would be well. Typically, though, those letters are much more excited about a wonderful new piece of physics Einstein has heard of. The German physicist Philipp Lenard had discovered that electrons could be knocked out of the surface of a metal by shining ultraviolet light onto it. Strangely, he had also discovered that the energy of the electrons ejected from the metal surface did not depend on how bright the light was. Whether it was faint or dim the electrons always came out with the same energy (essentially, the same speed). How could this be? Just four years later, Einstein would explain the phenomenon, and in 1922, he would receive the Nobel Prize for his explanation, with profound effects on the lives of himself and Mileva. But in 1901, the promise of the spring soon faded once again into uncertainty about their futures.

Einstein quite liked his brief time as a teacher, and was especially pleased to discover that after five or six hours spent at the task (schools started early in Switzerland) he was still fresh enough to work in the library or on interesting problems at home, so that he could continue his scientific work even if he didn't have any official connection with a university. And on Sundays he could take the train in to Zurich to see Mileva. But when the temporary job ended, and the reality of the implications of Mileva's pregnancy began to sink in, the future once again looked less promising, with no sign yet of the promised vacancy at the Bern Patent Office. In July, Mileva failed her exams again, almost as badly as she had done the year before. Abandoning all hope of a PhD, pregnant and depressed, she went home alone to see her parents and break the unwelcome news. Einstein took on an unsatisfactory job as a private tutor just to make ends meet and devoted all the time he could to writing a new dissertation, developing his ideas concerning the forces between molecules. He hoped to use this to obtain a PhD, which he now regarded as his best chance of getting an academic job.

In the autumn, Mileva retuned to Switzerland but couldn't stay with Einstein, or be seen with him, since her now visible pregnancy would have compromised his position as a respectable teacher. She stayed in a nearby town, while he made a succession of excuses that he could not find time to get away to see her. They had little contact, and once again she went back to her parents. In November, Einstein completed his dissertation and submitted it to the University of Zurich, scraping up the fee of 230 francs required for it to be considered. Professor Alfred Kleiner read the dissertation, but infuriated Einstein first by taking until well into the new year to comment on it, then advising the young man to withdraw it before it went any further. A furious Einstein did so (which at least meant he got his 230 francs back), and ranted to his friends about the incompetence of all professors, especially Kleiner. But the professor probably did Einstein a favour. No copy of the dissertation survives, and we cannot know what was wrong with it, but judging from the papers in *Annalen der Physik* on which it is supposed to have been based, Kleiner probably had a point. Einstein himself later described the paper as 'worthless'.[9]

Rescue

In any case, Einstein's anger was a little assuaged by news he received while Kleiner was still considering the dissertation. On 11 December he heard from Grossmann that the vacancy in the patent office was about to be advertised. Uniquely in the history of the patent office up to that time, the ad specified that applicants should have a university education with a 'specifically physical direction'; before 1902, there were no physicists in the Swiss Patent Office. Impulsively, as well as applying for the post, Einstein gave up his job and moved to Bern in January 1902, blissfully unconcerned that even if he got the job he would not be starting immediately. At about the same time, he became a father; after a difficult (indeed, life-threatening) labour Mileva gave birth to a little girl, Lieserl.

Although their correspondence clearly shows that the couple initially intended to keep their daughter, there was no prospect of Mileva bringing her to Switzerland. A respectable Swiss civil servant could not possibly be seen to be flouting conventional morality – and conventional morality in Bern at the time was so strict that women were legally forbidden to smoke in the street. There was no prospect of marriage until he had the patent office job – and there would be no prospect of the job if the authorities knew about his relationship with Mileva. So, in the short term at least, Einstein would be living as an impoverished but independent bachelor in Bern. It was almost an echo of his early days as a student in Zurich.

The Swiss bureaucracy only moved slowly towards the process of carrying out interviews and appointing new patent officers to fill the two vacancies that were becoming available, and meanwhile Einstein had to live. He decided to make a little money by private teaching, helping students at the University of Bern, and advertised his services in the local paper, picking up a couple of students willing to pay the modest fee he requested. One if these men, Louis Chavan, soon became a good friend. Einstein already had some contacts in Bern. An old schoolmate from Aarau, Hans Frösch, was now studying medicine at the University, and Paul Winteler, one of Marie's brothers, was studying law. Another old friend, Max Talmey, was travelling in northern Italy in the spring of 1902; after visiting Einstein's parents in Milan he called in to see Einstein himself in Bern. Talmey found Einstein living in a 'small, poorly furnished room' and struggling to make ends meet.[10] But if things were so hard financially that Einstein wasn't even getting enough to eat, socially and intellectually he was having the time of his life.

During the Easter vacation, Einstein ran the ad offering his services as a tutor again. This time, one of the responses came from Maurice Solovine, a 26-year-old

Romanian student from a wealthy family who was fascinated by the big ideas in physics and philosophy, but had no clear sense of what direction to follow. In him, Einstein found a kindred spirit, and the idea of Solovine paying for his education was soon forgotten as they became firm friends and met frequently to discuss the big issues of the day. Early in the summer, they were joined by Conrad Habicht, a mathematics student, and took to calling themselves the 'Olympia Academy', meeting regularly in the evenings and working their way steadily through books on the big ideas in philosophy and physics. Among the most influential of these works on Einstein himself were those by David Hume, Ernst Mach and Henri Poincaré. It was Poincaré who introduced Einstein to the concept of non-Euclidean geometries, mathematically self-consistent and logical versions of geometry in which, for example, the angles of a triangle do not add up to 180° and parallel lines can either meet or diverge from one another.* Poincaré's book *Science and Hypothesis*, which the 'Academy' studied and discussed in detail, poses an interesting question about the relationship between these geometries and the world we live in. Poincaré asked his readers what would happen if astronomers discovered that a pair of parallel light rays travelling through space eventually converged on one another. Would they conclude that space obeyed

---

* Coincidentally, Marcel Grossmann was studying non-Euclidean geometry for his PhD at around the same time.

non-Euclidian geometry? Or would they conclude that some unknown force was bending the light rays? Poincaré had no doubt that the answer would be in favour of the unknown force.

While Albert enjoyed himself in poverty in Bern, Mileva had returned to Zurich, without her baby, who was left with her relatives or a friend back home. Einstein's visits were rare, even when she moved to a small town closer to Bern, and in spite of the warmth of his letters to her it seems likely that the relationship might have ended in the way his relationship with Marie Winteler had ended, if it had not been for the child and the sense of duty that he felt towards the woman that he had, in what was still the language of the day, 'ruined'.

Mileva's prospects brightened in May 1902, when Einstein was at last called for an interview at the patent office and duly offered the post of 'Technical Expert III Class'. The other person appointed at the same time was an engineer, Heinrich Schenk. The patent office had recognised the need for a physicist because so much of their work now concerned inventions based on the application of electromagnetism, but Einstein's lack of experience in technical matters meant that he was initially appointed on probation, at a salary of 3,500 francs a year (roughly twice what he would have got in a university assistantship). Although he took up the post in June 1902, it wasn't until September 1904 that his appointment was confirmed as permanent and his salary increased to 3,900 francs.

But even the initial salary was a fortune to Einstein, who happily wrote to Mileva that now he could abandon the 'annoying business of starving'.[11]

There was now no financial impediment to the marriage. But there was still strong opposition from Einstein's parents, who knew about the baby. Einstein's mother Pauline, in particular, regarded Mileva as a scheming foreign hussy who was trying to trap her son. The situation wasn't helped by the fact that even though the Einsteins were secular Jews, nobody in the family had married a Gentile before. In spite of his independent nature and rebellious streak, Einstein couldn't bring himself to defy them on this occasion. His reluctance may have been partly due to a desire not to cause them more grief at a time when the family was already under stress. The latest business hadn't actually gone bust, but it was clear that it would never be profitable enough to repay the huge debts that the Einsteins owed to their relatives, who had begun pressing for the money. In addition, Einstein's father, Hermann, was ill with heart trouble, undoubtedly exacerbated because of the stress caused by his years as an unsuccessful businessman. On October 1902, his heart finally gave up the struggle; Hermann was just 55.

According to Einstein's sister Maja, on his deathbed Hermann relented and gave Einstein formal permission to marry Mileva.[12] Ironically, though, his death left Einstein rather less able to support a wife, since he began sending part of his salary to Pauline each month, to help her to pay off the debts her husband had left her with. Nevertheless, towards the end of the year Mileva moved to Bern, and on 6 January 1903 the couple were married at the register office. The only witnesses were the other two 'Olympians', Habicht and Solovine. Much later, to his biographer Carl Seelig, Einstein would admit that he had married from a 'sense of duty' in spite of his feelings of 'internal resistance'. For her part, Mileva still seems to have regarded it as a love match, and to have happily taken on the domestic role, cooking and looking after her new husband, while tolerating his friends. The only problem, from her point of view, was what to do about Lieserl.

The couple settled into a comfortable existence in their apartment in Bern. Albert wrote to his old friend Besso that: '[I] am living a very pleasant, cosy life with my wife. She takes excellent care of everything, cooks well, and is always cheerful.'[13] While Mileva wrote to her friend Helene Savic: '[Albert] is my only companion and society and I am happiest when he is beside me.'[14]

But she was not *his* 'only companion and society'. He still had his work at the patent office, which he enjoyed, and his friends, including the 'Olympians'. And he had his research, which was now developing along extremely promising lines, which he carried out in his own time and also when things were quiet in the patent office. Between 1902 and 1904 Einstein produced a series of three papers, all published in the *Annalen der Physik*, which came close to making him a name as an important member of the scientific community. Working completely on his own, he found how to interpret the laws of thermodynamics mathematically entirely in terms of the statistical behaviour of a myriad of tiny particles, laying the foundations of statistical mechanics. The only snag was, entirely unknown to Einstein, the American Josiah Willard Gibbs had beaten him to it and published what became the standard work on statistical mechanics in 1902. Their two approaches were almost the same, but Einstein only came across Gibbs's book in 1905, after it had been translated into German. The fact that two people independently came up with the foundations of statistical mechanics at about the same time is not surprising, since this step was (for anyone clever enough to take it) a logical progression from the work of the pioneers such as Maxwell and Boltzmann. It was just Einstein's bad luck that on this occasion he came second, but anyone who read and understood those papers in the *Annalen der Physik* can have been left in little doubt that the author was a scientist of note.

We don't know what discussions took place between Albert and Mileva in the first months of their marriage concerning Lieserl, but in August Mileva set off back to her homeland to sort the situation out. She was away for a month, during which she found out that she was pregnant again, and returned without the little girl. Historians have scoured the fragmentary correspondence that survives from the period, and the official records, in an effort to find out what happened to Lieserl, but in effect she just disappeared from sight in September 1903. The consensus is that she was adopted, most probably given up formally to strangers or possibly informally taken into the family of Mileva's friend Helene Savic and given another name. A reference to scarlet fever in one of Albert's letters to Mileva also raises the possibility that she died in infancy, as so many children did at the time. Whatever Lieserl's fate, when Mileva returned to Bern in the autumn of 1903 it was essentially a fresh start, crowned by the arrival of Einstein's first son, Hans Albert, on 14 May 1904.

By then, there were major changes in Einstein's circle of friends. Habicht and Solovine had finished their studies and moved on, but by that time Einstein had a more than adequate scientific sounding board to replace them. His friend Michele Besso had been finding it hard to make a living as a freelance engineer, and when a vacancy for 'Technical Expert II Class' came up at the patent office, Einstein drew Besso's attention to it. Besso applied and got the job, starting in the summer of 1904. He was a grade above Einstein and earned an annual salary of 4,800 francs (Einstein's raise to 3,900 francs would not come through until September), but this was entirely appropriate to his age and experience.

Now Einstein had a companion he could talk to about science during breaks at work, while walking to and from the patent office, and during his so-called leisure time. But there was very little real leisure time. Contemporary accounts describe how even when Einstein was pushing

Hans Albert about in his baby carriage on a Sunday afternoon stroll, he would have his pipe in his mouth and a notepad resting on the carriage, ready to write down his thoughts when inspiration struck. Thanks to Mileva, he had no domestic worries at all and never had to concern himself with routine details like cooking and cleaning. It was in this sense that she would make a major contribution to his miraculous year, and although she might have dreamed of sharing a scientific partnership with Albert like that of Marie and Pierre Curie (who won the Nobel Prize together in 1903), it was not to be.

With the breakup of the 'Olympians', the arrival of his son and Besso on the scene, and three solid scientific papers under his belt, Einstein now seems to have taken himself more seriously. Before, ideas had been tossed about in discussions over coffee and wine, but few of them had been followed through rigorously. Now, he buckled down to work through some of the ideas that had been floating around in his head for years. First, he would complete a new dissertation, making use of his growing understanding of statistical processes, and make a proper try for that elusive PhD. And then, there were three other ideas he had, concerning the reality of atoms, Planck's light quanta and the old puzzle about what the world would look like if you could travel at the speed of light. By the end of 1904 the scene was set, although even Einstein cannot have sensed it, for the greatest outpouring of scientific creativity since the time of Isaac Newton.

2 The *Annus Mirabilis*

The doctoral thesis; Jiggling atoms; Particles of light; The special one

In March 1905, Albert Einstein celebrated his 26th birthday; two months later, his son Hans Albert celebrated his first birthday. That year, Einstein was living in a settled household with a steady job, his close friend Besso to talk to about scientific matters, and a settled routine for working on his own scientific projects in his own time. This didn't leave much time for his family, and although he was a loving father he would hardly be regarded as a role model today. The 'problem' of Lieserl had been resolved, for good or bad, and once it was resolved his daughter seems to have been dismissed from his mind. Mileva, for all her one-time scientific aspirations, was essentially a conventional housewife, looking after her man so that he could concentrate on higher things. And as for being a father, rather than suffering the distractions most new fathers go through, Einstein just ignored them, the way he would ignore anything that threatened to disturb his scientific work throughout his life.* Several

---

* The story that in later life he gave up wearing socks so as not to have the bother of finding clean ones to wear is true, and typical.

accounts recall how on those occasions when he was supposed to be looking after the baby he might be found with his pipe in his mouth, rocking the cradle with one hand, while writing out calculations on his ubiquitous notepad with the other. This ability to switch off from the distractions of the world around him and to concentrate on the problems he was interested in was a major reason why he was able to produce such an outpouring of papers in 1905, and to maintain a high standard of scientific achievement for many more years to come.

The first hint of what Einstein had been up to in the early months of 1905 came in a letter he wrote at the end of May to his friend Conrad Habicht:

> I promise you four papers, the first ... deals with radiation and the energetic properties of light and is very revolutionary, as you will see ... The second paper is a determination of the true size of atoms by way of the diffusion and internal friction of diluted liquid solutions of neutral substances. The third proves that, on the assumption of the molecular theory of heat, particles of the order of magnitude of $\frac{1}{1000}$ millimeters suspended in liquids must already perform an observable disordered movement, caused by thermal motion. Movements of small inanimate suspended bodies have in fact been observed by the physiologists and called by them 'Brownian molecular movement'. The fourth

paper is at the draft stage and is an electrodynamics
of moving bodies, applying a modification of the
theory of space and time; the purely kinematic part
of this paper is certain to interest you.[1]

That has to be one of the most remarkable letters in the
history of science. The first paper Einstein mentions
established the reality of light quanta (what we now call
'photons'), and was so revolutionary that it earned him
the Nobel Prize – although because it was so revolutionary it took sixteen years for the rest of the scientific
community to catch up with him and make that award.
The second paper formed the basis of his doctoral dissertation. The third proved the reality of atoms. And
the fourth presented a bemused world with the Special
Theory of Relativity.

The doctoral thesis

I will describe these pieces of work in a slightly different
order, starting with the doctoral thesis. One good reason
for doing this is that this was the least revolutionary of
the four papers, and Einstein knew it. When he decided
to make a second attempt at obtaining a PhD from the
University of Zurich, Einstein didn't set out on a specific
new project to achieve that goal, but simply seems to
have looked at the various projects he had in hand and
chosen the most straightforward one, based on solid, traditional methods, that wouldn't tax the imaginations of

the professors at the university too much. He was also careful to choose a piece of work based on experimental observations, although he didn't carry out the experiments himself. There is a story, originating with Einstein's sister Maja,[2] that Einstein first offered his paper on the electrodynamics of moving bodies (the Special Theory of Relativity) and it was rejected because the examiners didn't understand it; but this seems to be a myth. There is no evidence of such a rejected application, Einstein had more sense than to baffle the examiners in this way, and in any case the paper was the last of the four to be completed, as the letter to Habicht shows.

Einstein actually completed the paper that became his dissertation, prosaically titled 'A New Determination of Molecular Dimensions', at the end of April, but he didn't submit it to the University of Zurich until 20 July 1905. The delay may have been because he was busy working on his other ideas in the spring and didn't make a final decision about which paper to submit until then. Although the title of the dissertation focuses on the sizes of molecules, the technique Einstein describes actually also gives a measurement of the number of molecules (or atoms) present, in this case in a solution. This is typical of the kinds of methods used to estimate the numbers and sizes of atoms and molecules in the 19th and early 20th centuries, and shows Einstein building on what has gone before, rather than leaping off in a new direction.

The problem was that in mathematical descriptions of the behaviour of atoms and molecules, both the sizes and the numbers of these particles appear in the equations. As we all learned in school, if you have one equation involving one unknown quantity (usually denoted by x), you can solve the equation to find a value for the unknown quantity. But if you have a single equation involving two unknowns (x and y), you cannot find out what the values of the unknowns are. To do that, you need two different equations each involving x and y. So all the 'classical' methods for determining the sizes of molecules and the numbers of molecules in a certain amount of matter depended on using two equations to work out the two unknowns.

The number that comes into these calculations was called Avogadro's number, after the Italian who came up with the idea in 1811.* It is the number of particles (atoms or molecules) contained in an amount of material whose weight in grams is numerically equal to the atomic (or molecular) weight of the substance. The atomic weight of carbon, for example, is twelve; so twelve grams of carbon contains Avogadro's number of atoms. The atomic weight of hydrogen is one, but each hydrogen molecule contains two atoms, so its molecular weight is two. So two grams of hydrogen gas also contains Avogadro's number of molecules – and so on.

---

* This number is called Avogadro's 'constant' today, but we shall stick with the name familiar in Einstein's day.

One early attempt to work out the value of this number, and simultaneously the sizes of molecules, was made by the German Johann Loschmidt in the mid-1860s. His calculations involved the average distance travelled by particles in a gas between collisions with one another (called the 'mean free path') and the fraction of the volume of the gas actually occupied by the molecules themselves. He reasoned that in a liquid all the molecules must be touching each other with no gaps in between; so measuring the density of the liquid, which depends on the number and size of the molecules present, would tell you the volume occupied by Avogadro's number of molecules. When the liquid is heated to become a gas, the actual molecules must still occupy the same volume as the original liquid, but now with lots of empty space between them. It's only in the gas that the mean free path comes into the calculations.

Loschmidt carried out his calculations for air, which is almost completely a mixture of oxygen and nitrogen, and had to use estimates for the density of liquid nitrogen and liquid oxygen which were not as accurate as modern measurements. He combined these with calculations of the mean free path, which also depends on the number and size of the molecules present, based on measurements of the way the pressure exerted by air changes when it is squeezed into a smaller volume. He found that a typical molecule of air must be a few millionths of a millimetre across, and estimated Avogadro's number to be 0.5×10^{23}

– which means a 5 followed by 22 zeroes, or 50 thousand billion billion.

Einstein's approach to the problem was in the same spirit of solving two different equations simultaneously, but used a very different kind of physical system. He realised that the sizes of molecules (and Avogadro's number) could be inferred from measurements that had already been carried out on the behaviour of solutions of sugar in water. But, as we have said, he didn't do the experiments. What was new about Einstein's work (and what justified the award of his PhD) was the mathematical way in which he calculated how molecules of sugar would behave in such a solution, and how this would affect the measurable properties of the solution. What was particularly clever about the work wasn't that it gave a value for Avogadro's number or the size of molecules – the techniques based on the kinetic theory of gases, such as Loschmidt's method, had already done that. Einstein's special contribution was to find a way to get results as good as those obtained from the kinetic theory of gases using liquids alone. Previously, estimates based on studies of liquids had been very rough and ready. Along the way, as we shall see, he developed techniques with widespread applications for industry wherever suspensions of particles in liquids are used.

The technique depended on the fact that sugar molecules are very much larger than molecules of water. In fact, as Einstein realised, because some water molecules actually attach themselves to the sugar molecules in the solution, the effective size of the sugar molecules is even bigger, which makes the assumptions used in his calculations even more accurate.

It is easy to describe the thinking behind those calculations. When something is dissolved in water, the viscosity of the solution – its stickiness – increases. By assuming that each sugar molecule is a large sphere embedded in a sea of much smaller water molecules, Einstein was able to work out an equation which related this change in viscosity (which can be measured) to the total volume of the fluid occupied by the sugar (which depends on two unknown quantities, the size of each sugar molecule and the number of sugar molecules present). Because experimenters always measure the weight of sugar (or other stuff) being added to the solution, the number of molecules present can always be expressed in terms of Avogadro's number.

Then, Einstein looked at the way sugar diffuses through water, and calculated the force acting on a single sugar molecule as it moves through the sea of water molecules. This could be related to another measurable property of the solution, called its 'osmotic pressure', through another equation which itself depended on both the number of sugar molecules present and their size. So Einstein derived two equations, each of which included the two unknown quantities, Avogadro's number and the size of a sugar molecule, and each of which was directly related to measurable properties of the solution, its viscosity and the osmotic pressure.

Once he had that pair of equations, it was a simple matter to plug in the numbers for viscosity and osmotic pressure that were already well known and had been published in standard tables listing the properties of such solutions. For the record, the 'answers' that came out of the equations were that sugar molecules (which are much bigger than molecules of nitrogen or oxygen) have a radius of about 9.9×10^{-8} cm (9.9 hundred-millionths of a centimetre) and Avogadro's number is 2.1×10^{23} (210 thousand billion billion). This pretty much agrees with Loschmidt's estimate, but the impressive part of the dissertation is the bit we can't put into words, the sophisticated mathematical techniques that Einstein used to deduce the relevant equations.*

The best way to appreciate just how good the maths was is to look at what the professors who examined the work said in their official report to the University of Zurich. Alfred Kleiner commented that: 'the arguments and calculations to be carried out are among the most difficult in hydrodynamics and could be approached only by someone who possesses understanding and talent for the treatment of mathematical and physical problems ... Herr Einstein has provided evidence that he is capable of occupying himself successfully with scientific problems'.

---

* Remember that what matters is not so much the number on the front, 0.5 or 2.1, as the agreement of the number of 'powers of ten' in the exponent, 23.

His colleague Heinrich Burkhardt said that: 'the mode of treatment demonstrates fundamental mastery of the relevant mathematical methods.'[3] Einstein himself later told his biographer Carl Seelig that the only official comment he had received on the dissertation was that it was too short, and that in response he had added a single sentence, whereupon it was accepted. Even though the dissertation itself was officially accepted by the University of Zurich early in August 1905, it took until 15 January 1906 to complete all the various formalities required by the University for the degree to be conferred. So all the papers he completed during the *annus mirabilis* were the work of simple 'Herr Einstein', not 'Herr Doktor Einstein'.

Just after the thesis was accepted, Einstein submitted a slightly revised version to the *Annalen der Physik*, but publication was delayed until 1906 because the editor of the journal, Paul Drude, knew of some more accurate and up-to-date measurements of the properties of sugar solutions than the ones Einstein had used. When he asked Einstein to take account of these data, the result was a slight change in the numbers he came up with – in the right direction, we now know. Even that wasn't the end of the story, because Einstein's paper eventually encouraged other experimenters to measure the relevant properties of these and other solutions even more accurately. It also turned out that Einstein had made a minor error in his calculations, which had not been spotted by either of his examiners or by Drude. The final, definitive version of Einstein's method for calculating

Avogadro's number from the properties of sugar solutions only appeared in the *Annalen der Physik* in 1911, and gave a value of 6.56×10^{23}, which is very close to the accepted modern value, 6.02×10^{23}.

The saga of Einstein's doctoral dissertation did not end there, however. In scientific terms, this is the most mundane of the papers he wrote during the *annus mirabilis*. But it has one curious distinction. It became far more widely quoted than any of his truly revolutionary papers from the same year.

One way in which scientists measure the value of scientific papers is to record how often they are referred to in other scientific papers. This is by no means a perfect system, as witnessed by the fact that Einstein's original paper on the Special Theory of Relativity has been very seldom referred to. The reason for that, of course, is that the content of the paper quickly became part of the established fabric of science, something taught from textbooks that 'everybody knows', so that few scientists have even read the paper, let alone cited it.

By contrast, the paper based on the doctoral dissertation has been very widely cited. Just how widely was brought home in 1979, as part of the celebrations to mark the centenary of Einstein's birth. Two researchers carried out a survey of the citations received not just by Einstein's papers but by all the papers in science (what they called the 'exact' sciences, like physics and chemistry) published before 1912.[4] The twist was that they only looked at citations in papers that had themselves been published between 1961 and 1975; so they came up with a list of all the papers that were still important enough to be quoted at least 50 years after they had originally been published. Out of the top eleven 'most cited' papers in this survey, four were by Einstein. (No other scientist had more than one paper in the top eleven.) And top of the four papers by Einstein came his doctoral dissertation.

Why was this seemingly mundane paper cited so often between 1961 and 1975? Simply because it is mundane. It deals with practical things important in the everyday world – the behaviour of fluids with particles suspended in them. The equations Einstein derived are relevant to (among other places) the dairy industry, where it is important to understand and predict the behaviour of milk during the process of making cheese; the study of pollution and the way tiny particles called aerosols get spread through the atmosphere; and problems involving the behaviour of cement being transported in liquid form, and the design of the lorries to carry the cement. The work Einstein did for his doctorate in 1905 turned out to be of widespread importance in many practical applications in the second half of the 20th century, and is still relevant today, more than a hundred years after the dissertation was written.

Jiggling atoms

For the same reasons, the second most cited of Einstein's papers in that 1979 survey did not concern the Special

Theory of Relativity or quantum physics (indeed, neither of those papers even made the top four), but the phenomenon known as Brownian motion. Appropriately, it was the paper on Brownian motion that Einstein returned to as soon as he had finished drafting what would become his doctoral dissertation, at the end of April 1905. On 11 May, Paul Drude received a paper with the splendid title 'On the Motion of Small Particles Suspended in Liquids at Rest Required by the Molecular-Kinetic Theory of Heat'.* He had no hesitation in accepting it for publication.

Brownian motion got its name from the Scottish botanist Robert Brown, who first studied the phenomenon in detail in 1827. Intriguingly, though, Einstein wasn't trying to explain Brownian motion in this paper; indeed, in a sense he was predicting it, on the basis of his statistical approach to the kinetic theory, honed in the series of three papers mentioned earlier. That's why the term 'Brownian motion' doesn't appear in the title of the paper. In the first paragraph of the paper, Einstein says:

> It is possible that the motions to be discussed here are identical with so-called Brownian molecular motion; however, the data available to me on the

---

* You sometimes see slightly different versions of the titles of Einstein's papers. The titles and quotes from Einstein's papers that are used here, are taken from Stachel, which is the most accessible source.

latter are so imprecise that I could not form a judgement on the question.

But since we now know enough to form that judgement, it
makes sense to introduce this aspect of Einstein's work by
looking at just what it was that Robert Brown discovered.

Even before Brown's time, people had noticed the
way tiny grains of material, notably pollen, seem to dance
about in a jittery kind of motion, something like running
on the spot, when they are suspended in a liquid such as
water and observed through the microscope. So Brown
didn't discover Brownian motion. Before Brown's work,
however, the obvious explanation for this motion seemed
to be that the particles were alive – after all, pollen grains
are a kind of plant equivalent to the sperm cells in ani-
mals, and if sperm can move under their own steam, why
shouldn't pollen? When Brown began his detailed studies in the summer of 1827 (the results were published in
1828), he thought that this was the most probable explanation. But then he made the next logical step. He took
a series of clearly inanimate materials, such as ground
up fragments of glass and granite, and suspended them
in water. He found exactly the same behaviour for these
definitely non-living materials, proving that the motion
of a particle in suspension has nothing to do with any
mysterious life force. 'These motions,' he wrote in that
1828 paper, 'were such as to satisfy me, after frequently
repeated observation, that they arose neither from

currents in the fluid nor from its gradual evaporation, but belonged to the particle itself.'[5] It was this discovery, a result of the truly scientific way he went about his work, that meant his name would be forever associated with the phenomenon.

But if a life force wasn't causing the motion, what was? Over the next few decades, people considered the possibility that convection currents might be involved (in spite of Brown's comments), or electrical effects, or the same force that caused capillary action, and other more or less wild ideas. The key experimental discoveries were that the speed of this jiggling increased if the temperature of the water increased, and was less for bigger particles. Combining this with the ideas of the kinetic theory gave rise to the suggestion that the particles were being bombarded by the molecules in the water, and were being jerked about in response to the kicks they received from individual molecules. But in order for a single molecule to produce a visible shift in a pollen grain or a speck of granite dust, the molecule would either have to be impossibly big, or travelling impossibly fast.

This was more or less where the puzzle of Brownian motion stood at the beginning of the 20th century. It is clear from his writings, though, that Einstein had not read up on all of these developments, and was not up-to-date on the subject. He was aware of the phenomenon of Brownian motion, but his theoretical studies of how particles suspended in liquids ought to move were not specifically intended to explain that phenomenon. Rather, they were a logical development from the work in his doctoral dissertation.

As Einstein has told us, what he was really interested in at that time was proving the reality of atoms and molecules. He was completely convinced of the validity of the kinetic theory of heat, and saw in this extension of his PhD work a way to convince others as well. For this purpose, the distinction between atoms and molecules is of no significance. *Atoms* are the fundamental component of elements, such as hydrogen and oxygen, and *molecules* are the basic components of compound substances, such as water (where two hydrogen atoms combine with one oxygen atom in each molecule of water).

At its simplest, the kinetic theory says that everything is made of tiny particles (atoms or molecules) which can be regarded as little, hard spheres. In a solid, the little spheres are packed closely together and do not move past one another. In a liquid, the little spheres buffet each other and slide past one another like people moving through a dense crowd, but they are still essentially touching all their neighbours. In a gas, the little spheres fly freely through empty space, bouncing off each other and the walls of any container they are in. The hotter a substance is, the faster the spheres move, which explains the transition from solid to liquid to gas as a substance is heated, and from gas back to liquid and then solid when it cools.

In the paper that became his doctoral dissertation,

Einstein had already used the idea that molecules of sugar dissolved in water are being bombarded by water molecules from all sides, and that the way the sugar molecules move through the sea of water molecules affects measurable properties of the solution: its viscosity and its osmotic pressure. The success of Einstein's results from that paper already provided powerful circumstantial evidence in favour of the kinetic theory, but even that was not direct proof that atoms and molecules exist. To obtain that, the effects of the bombardment by water molecules had to be scaled up somehow, to become visible, at least under the microscope. A pollen grain, tiny though it is by any human standard (about a thousandth of a millimetre across), is enormously much bigger than a water molecule (measured in millionths of a millimetre), or even a molecule of sugar. But Einstein made the huge mental leap of realising that as far as the behaviour of particles suspended in liquids was concerned, this was the only difference that mattered between a pollen grain (or a fragment of granite) and a sugar molecule. In what I shall refer to as the Brownian motion paper, he said:

> According to [the kinetic theory], a dissolved molecule differs from a suspended body only in size, and it is difficult to see why suspended bodies should not produce the same osmotic pressure as an equal number of dissolved molecules. We have to assume that the suspended bodies perform an irregular, albeit

very slow, motion in the liquid due to the liquid's molecular motion.

And he went on to calculate both that osmotic pressure and the nature of that irregular motion.

The osmotic pressure that comes into both this work and the doctoral dissertation is a curious phenomenon worth describing in a little more detail. If you have a container of water (or some other liquid), like a fish tank, it can be divided into two by putting in a barrier that has tiny holes in it, just big enough for water molecules to pass through. If you do this, the water can get from one side (either side) of the barrier to the other. This is known as osmosis. But if you now dissolve something (such as sugar) in the water on one side of the tank, the dissolved molecules are too big to get through the holes. The barrier in such a setup is then called a 'semi-permeable membrane', because it lets some molecules through but not others.* This is where things get interesting.

You now have a solution of sugar in water on one side of the membrane, and pure water on the other side. The result is a pressure that moves water molecules from one side of the membrane to the other. You might guess (most people do, the first time they come across this) that the

---

* All of this works for other solutions as well, of course, but we shall stick with sugar and water because that was the example Einstein used in his dissertation.

presence of the sugar molecules pushes water out of that half of the tank, making the solution stronger and raising the level of the water on the other side of the barrier. In fact, just the opposite happens. Water from the pure side of the tank passes through the membrane, making the sugar solution more dilute, and increasing the height of the liquid on the side of the barrier where the sugar is. The process only stops when the pressure of the extra height of liquid (the osmotic pressure) is enough to stop the flow of water molecules through the membrane.

This counter-intuitive behaviour is an example of the famous second law of thermodynamics at work. I don't have space to go into all the details here, but the relevant point, at the heart of that law, is that natural processes tend to even out irregularities in the Universe. On a grand scale, the Sun and stars are pouring out heat into the cold Universe; on a more homely scale, an ice cube in a glass of water melts, evening things out to produce an amorphous liquid. In the example of osmotic pressure, the water molecules that move into the sugar solution make the solution more dilute, more like the pure water, so that there is less contrast between the fluids on opposite sides of the semi-permeable membrane.

In the Brownian motion paper, Einstein first covered some similar ground to parts of his doctoral dissertation, but using a different (and rather more elegant) mathematical approach. His calculations involved the relationship between osmotic pressure, viscosity and the way individual particles suspended in the liquid diffuse through the sea of molecules surrounding them. But this time, he was describing the behaviour of particles big enough to see under the microscope.

The way Einstein set about his work, though, was just as important as the results he obtained. He realised that the kick produced by a single molecule hitting a particle as large as a pollen grain could not produce a measurable shift in the position of the large particle. But the large particle is constantly being bombarded by molecules, from all sides. On average, the kicks from one side are balanced by the kicks from the opposite side, so you might not expect the large particle to move at all. Einstein realised, however, that the important words are 'on average'. If you take a very small time interval, then just by chance at that instant the particle will be receiving more kicks on one side and fewer on another. The combined effect will be to shift the particle by a minute amount in the direction of least resistance. Then, in the next instant the pattern will change, and the particle will shift in another direction, and so on. Einstein's special insight into the nature of this kind of statistical fluctuation was that what happens during each of these small time intervals is entirely independent of what happens in any other time interval, even the one just before the one being considered.

Because of this independence, and the statistical nature of the fluctuations, the particle doesn't simply move to and fro around the same spot that it started from.

Nor does it keep moving in one direction. Einstein discovered that it gradually moves further and further away from its starting point, but following a zigzag path that has become known as a 'random walk'. He showed that wherever the particle starts from, the distance it moves away from its starting point depends on the square root of the time that has passed. So if it moves a certain distance in one second, it will move twice as far in four seconds (because 2 is the square root of 4), four times as far in sixteen seconds, and so on. But it doesn't keep going in the same direction – after four seconds it will be twice as far away from the start as it was after one second, but in a random and unpredictable direction.

This is called a 'root mean square' displacement, and the equation Einstein worked out for the displacement involves the temperature of the liquid, its viscosity, the radius of the particle and Avogadro's number. He used this equation and a value for Avogadro's number inferred from other experiments to predict that a particle with a diameter of 0.001 mm in water at a temperature of 17°C would shift a distance of 6 millionths of a metre from its starting point in one minute. But he also realised that if the predicted displacement could be measured accurately enough, the same equation could be used the other way around, to give a value for Avogadro's number.

The prediction provided a classic example of the scientific method at work, since measuring the way a particle moved away from its starting point would answer the question of whether the theory it was based on was right or wrong. As Einstein put it in his paper:

> If the prediction of this motion were to be proved wrong, this fact would provide a far-reaching argument against the molecular-kinetic conception of heat.

It wasn't proved wrong. Although Einstein didn't know it when he wrote the paper, in the early 1900s microscopists were already developing improved instruments, known as ultramicroscopes, that would be able to measure the kind of motion he was describing accurately enough to test the prediction.

The Brownian motion paper was published in July 1905, and almost immediately Henry Siedentopf, a German working with the new ultramicroscope, wrote to Einstein to tell him that the kind of motion described in his paper almost certainly was Brownian motion. It still wasn't possible at that time to test Einstein's detailed predictions, but he was sufficiently encouraged to write another paper, this time plainly titled 'On the Theory of the Brownian Motion', which he sent off to the *Annalen der Physik* in December; it was published in 1906. In this paper he developed his ideas further and also predicted that particles suspended in a liquid would experience a rotary movement, dubbed 'Brownian rotation', although he did not expect this to be observable.

It was extremely difficult to make the observations required with enough accuracy to test Einstein's predictions, and several researchers tried and failed over the next couple of years; but in 1908 the French physicist Jean Baptiste Perrin finally succeeded. Instead of trying to measure the displacement of tiny individual particles from their starting point, Perrin used another result that had by then emerged from Einstein's theoretical model, which predicted the way particles suspended in a solution would be arranged vertically.

In such a suspension, the particles would be tugged downward by gravity, gradually sinking to the bottom of the liquid. But superimposed on this very slow downward drift would be Einstein's random walk. The overall effect would be a vertical distribution of particles, with more at the bottom and fewer at the top, obeying a precise mathematical law (a specific exponential, decreasing with increasing height). Perrin's results exactly matched the predictions from Einstein's theory, and he even went one better by measuring the Brownian rotation that Einstein had predicted before anyone had seen it. He also used the observations to make an accurate measurement of the value of Avogadro's number.

The whole package finally established the reality of atoms and molecules, and the validity of the kinetic theory, silencing the few (by then, very few) remaining doubters. This work was so important that Perrin received the Nobel Prize 'in particular for his discovery of the equilibrium of sedimentation', as the citation put it, in 1926.

But Einstein's work had even wider applications. The kind of statistical methods he used, coupled with the idea of random events occurring independently in individual tiny intervals of time, proved fruitful across a whole range of topics in physics. In his second paper on Brownian motion (still written in, though not published in, the *annus mirabilis*), Einstein had pointed out the possibility of applying the same approach to the study of fluctuations in electric circuits (the phenomenon now known as 'noise'), and in years to come the technique would be widely applied in the new field of quantum physics. There, for example, exactly the same combination of statistical effects and random changes occurring in independent time intervals leads to an understanding of the nature of the half-life associated with radioactive processes.

Particles of light

This link with quantum physics is particularly appropriate, because the next paper I shall discuss from Einstein's miraculous year saw him laying one of the foundation stones on which the whole edifice of quantum theory was built. Curiously, although this was the one paper that Einstein himself referred to as 'very revolutionary' (in that letter to Conrad Habicht), in many ways it builds from his other work, and to modern eyes looks less revolutionary than the work on the Special Theory of Relativity.

But that is because we have become used to the idea that light exists in the form of tiny particles, called photons. In 1905, that really was a revolutionary concept – although it wasn't exactly new.

Isaac Newton thought of light as a stream of tiny particles, and used this model in his attempts to explain his observations of the way light is bent when it passes through a prism, how it is reflected by mirrors and why it can be broken up into all the colours of the rainbow, forming a spectrum. His 17th-century Dutch contemporary, Christiaan Huygens, had argued for a different interpretation of the same phenomena, based on the idea that light is a form of wave; but Newton's model held sway (largely because of the god-like status that his successors gave to Newton) until the work of the Englishman Thomas Young and the Frenchman Augustin Fresnel early in the 19th century. Although their contributions were equally important, what became regarded as the definitive proof that light travels in the form of a wave, like ripples on a pond, comes from what has become known as Young's double-slit experiment.

The experiment is based on shining light of a single colour (this would later be interpreted as meaning light of a single wavelength) coming from a light source through two holes in a screen. These could be two parallel thin slits, made with a razor, or two pinholes. We have all seen the pattern of circular ripples that spreads out from a point when a pebble is dropped into still water, and the more complicated pattern of ripples that is produced if two pebbles are dropped into the water simultaneously. The complications in the second pattern are caused by two sets of ripples interacting with one another – as physicists put it, 'interfering' with one another. The experiments showed that light spreads out from the holes (or slits) in Young's experiment in just the same way, and that two sets of waves, one from each hole, are interfering with one another. Young (and then many other people) proved this by placing a second screen on the other side of the two slits from the light source, and looking at the pattern of bright and dark stripes made on the second screen. Bright stripes are places where the two sources of light combine with one another to make an extra high wave, and dark stripes are places where the two sets of ripples cancel each other out, with one going up while the other goes down. This is quite different from the pattern that would be expected if light travelled in the form of tiny particles, like little bullets. The experiments are so precise that the spacing of the stripes on the second screen can be used to calculate the wavelength of the light involved – proof, indeed, that light travels as a wave.

It's worth putting this dramatic discovery in its historical perspective. Newton had spelled out his ideas about light and colour in a great book, *Opticks*, published in 1704. His image of light as a stream of tiny particles held sway for almost exactly 100 years, until the work of Thomas Young at the beginning of the 19th century.

That's almost the same as the time interval from Einstein's work in 1905 to the present day – and the interval from Young to Einstein is the same as the span from Newton to Young. To suggest at the beginning of the 19th century that Newton had made a major blunder was very much as if evidence were uncovered today showing that Einstein had made a major blunder. The discovery was dramatic, and it took time for people to be convinced. But in the 100 years from Young to Einstein, a great deal more evidence did come in to show that light travels as a wave.

We have already mentioned the most important piece of that evidence – James Clerk Maxwell's discovery of the equations that describe how electromagnetic waves (or 'vibrations of the ether', as they would have put it then) move through space. Maxwell's equations describe waves, and they predict the speed with which those waves move. This speed is exactly the same as the speed of light. What more proof could be needed that light travels as a wave? By 1900, the idea that light, and other forms of electromagnetic radiation, existed in the form of waves seemed as solid a foundation of science as the idea that apples fall downwards from trees. But then the first crack appeared in this foundation.

Two of the big areas of scientific interest in the middle and late 19th century were thermodynamics (which dealt with energy) and light (which had been identified as a wave and was also a form of radiant energy) – 'light' in this context refers to all kinds of electromagnetic radiation, including invisible infrared heat, radio waves and ultraviolet light. It was clear that there is a relationship between heat (energy) and light. A piece of iron that is just warm to the touch doesn't radiate any visible light at all, but as it is heated further it glows first red, then orange, then white as its temperature increases. Indeed, the relationship between colour and temperature was so well known in a qualitative way that in the days before accurate scientific measurements were possible, potters used to gauge the temperature of their kilns by looking at the colour of the pots they were firing. But what was the precise, quantitative relationship between light and energy? What were the equations that could describe, or predict, the colour of a hot object from its temperature alone?

The first person to tackle this puzzle in a quantitative way was the German physicist Gustav Robert Kirchoff, at the beginning of the 1860s. Kirchoff was especially interested in spectroscopy, studying the distinctive patterns of lines in a spectrum (looking not unlike a modern barcode) corresponding to different elements. But he also developed a thermodynamic approach to understanding the relationship between light and energy through his idea of a 'black body'. A black body would be an object which absorbed entirely all the radiation that fell on it – a perfect absorber. Of course, it was impossible for experimenters to make a perfect black body to study in the lab, but Kirchoff came up with a very close approximation.

He devised an experiment involving a closed container painted black inside, with a tiny pinhole as the only opening to its interior. Any radiation that entered through the pinhole would be absorbed, and only a tiny amount of radiation would escape through the pinhole, making it very nearly a perfect black body.

But this was only the first step. According to the thermodynamic rules, an object that absorbed all kinds of radiation should also radiate all kinds of radiation, with no complications involving things like spectral lines. If the container were heated, the glowing walls inside would produce light which would bounce around inside and get thoroughly mixed up before escaping from the pinhole, in the form that became known as 'cavity radiation', or (more commonly today) 'black-body radiation'. According to the thermodynamic principles, this would be a pure form of light, with a colour that depended only on the temperature of the container – the black body. Crucially, the colour of the radiation did not depend on what the container was made of.

But the black-body radiation is not light of a pure single colour. It is always a mixture of different colours – that is, different wavelengths – of light. For any particular temperature, however, there is always more energy radiated in one group of wavelengths, with less energy radiated at both longer and shorter wavelengths. As the cavity is heated, the peak intensity of the light shifts from the longer wavelength end of the spectrum (red) through the familiar colours of the rainbow (orange, yellow, green, blue and so on). So a red-hot piece of iron (or anything else) radiates mostly red light, but also some infrared radiation (at longer wavelengths) and some yellow and orange light (at shorter wavelengths).*

A graph representing the spectrum of a black body radiating in this way looks like a little hill, with a peak at a particular wavelength corresponding to the temperature of the black body, and slopes rolling down on either side. Even without any understanding of why this should be so, the discovery had immediate practical uses. For example, the shape of the spectrum of the Sun very closely follows this 'black-body curve' for an object with a temperature of about 6,000°C; so astronomers could measure the temperature of the surface of the Sun (and, indeed, of other stars) without ever leaving the Earth. What was needed to complete Kirchoff's work, though, was to find an equation that described the shape of this hill, and a physical basis for that equation. That proved extremely difficult, and Kirchoff, who died in 1887, didn't live to see it.

When he died, Kirchoff was the professor of Physics at the University of Berlin. His successor, who wasn't appointed until 1889, was a 31-year-old physicist of the old school, Max Planck. Planck was so conservative, in

---

* A white-hot object looks white because the peak of its spectrum is in the middle of the rainbow of colours; so it radiates all the colours, which combine to make white light. If it were even hotter, it would look blue, as some stars do.

scientific terms, that he might almost be described as a reactionary. He hated the way ideas involving probability and statistics, rather than the certainty of conventional mathematical equations, were being introduced into physics by people like Boltzmann, and had yet to be convinced of the reality of atoms. In 1882 he had stated dogmatically that: 'despite the great success that the atomic theory has so far enjoyed, ultimately it will have to be abandoned in favour of the assumption of continuous matter.'[6] What stuck in Planck's throat was the idea that matter could come in discrete lumps, with gaps in between. He much preferred the image of electromagnetic radiation as smooth and continuous waves, and expected that matter would also be found to be smooth and continuous, regardless of how successful the kinetic theory and Boltzmann's ideas on thermodynamics might seem to be.

In 1897, though, J.J. Thomson, working at the Cavendish Laboratory in England, showed that the streams of radiation known as cathode rays were actually made up of tiny, electrically-charged particles, which soon came to be known as 'electrons'. Whatever the reality of atoms, there could be no doubt that matter contained these little particles, and since they carry electric charge it seemed clear that there must be some connection between the behaviour of electrons in matter and the way matter radiated light. In particular, Maxwell's equations told physicists that a charged particle vibrating to and fro (an electric oscillator) must radiate electromagnetic waves.

By the end of the 19th century, though, the problem of how to describe black-body radiation mathematically had run in to a cul-de-sac – or rather, two cul-de-sacs. In the early 1890s, Wilhelm Wien, a lecturer at the University of Berlin, had come up with a mathematical description of black-body radiation that produced a graph which exactly matched one side of the spectrum, the short wavelength side of the hill, but was hopelessly wrong at describing the shape of the curve for longer wavelengths. In 1900, the English physicist Lord Rayleigh found another equation, based on different physical assumptions, that predicted a curve which exactly matched the black-body curve on the long wavelength side of the hill, but was hopeless when applied to the shorter wavelength side.* This discrepancy clearly showed that there was some fundamental misunderstanding of the nature of black-body radiation, and caused consternation among physicists. But in the same year, 1900, Planck, who had been studying the problem of black-body radiation intensively since 1897, came up with something that looked like it might be the answer to the puzzle.

In October 1900, Planck came up with a formula – an equation – that described the entire black-body curve accurately, smoothing over the join between Wien's law and Rayleigh's law. This was essentially an

---

* His equation was later refined by James Jeans and became known as the Rayleigh-Jeans law.

empirical result, one worked out by trial and error, and it included two constants, one of which became known as Boltzmann's constant and the other of which, given the label h, became known as Planck's constant. Boltzmann's constant usually turns up in connection with the properties of gases, but, unlike the case of the constant c in Maxwell's equations, there was no obvious physical interpretation of the new constant h. Planck presented his results to a meeting of the Berlin Physical Society on 14 October, even though he was well aware that he still lacked any physical basis for the equation he had come up with. But he kept beavering away at the problem, and before the end of the year he had found a physical basis for the equation. It wasn't particularly palatable to him, but it was the best he could do.

Planck was working on the assumption that electromagnetic radiation was emitted or absorbed by matter because of the presence of the relatively newly discovered electrons jiggling about inside the matter. In trying to explain the nature of the cavity radiation studied by Kirchoff and others, he had to think of the walls of the cavity containing a large number of 'harmonic oscillators', each corresponding to a jiggling electron. Radiation would be radiated and absorbed, re-radiated and re-absorbed, re-re-radiated and re-re-absorbed, in a repeating process mixing up all the radiation to achieve a state of dynamic equilibrium, with the maximum amount of disorder, before it could escape from the pinhole.

Unfortunately for Planck, this kind of equilibrium was described by the rules of thermodynamics – indeed, it is known as thermodynamic equilibrium – and describing it mathematically involved some of the statistical techniques developed by Boltzmann (which is where his constant comes in). Even more unfortunately, there was no getting away from the fact that electrons were individual particles, not a continuum, no matter how much Planck might like the idea that matter was continuous. The only way to take account of their collective behaviour was, once again, to use the statistical techniques developed by Boltzmann and Maxwell in connection with the study of the behaviour of large numbers of atoms and molecules. In particular, Planck was forced to use these statistical methods to calculate the property of the array of harmonic oscillators known as 'entropy'. It's worth elaborating a little on this, since it would also be central to Einstein's work.

Entropy is a very important concept in thermodynamics, but it can be understood very simply as a measure of the amount of disorder in a system. The entropy of an isolated system (which means anything left to its own devices, with no constructive input of energy from outside) always increases, which is a scientific way of saying that things wear out. If you build a house and leave it untended for a few hundred years, it will crumble away; but if you put a pile of bricks in a heap on the ground and leave them alone they will never spontaneously arrange themselves into a house. People, and other living things, can hold the increase of entropy at bay for a time by making use of the energy in the food we eat (which ultimately comes from the Sun and is stored in plants), but in the very long term everything wears out.

Disordered systems have more entropy than ordered systems. That's why an ice cube placed in a glass of water melts, evening out the difference between the water and the ice, and why water diffuses the 'wrong way' through a semi-permeable membrane. It's why the radiation inside one of Kirchoff's cavities gets scrambled up into a complete mess. A chess board painted in black and white squares is an ordered system with relatively low entropy, but the same piece of board painted with the same amount of black and white paint but mixed to a uniform grey colour has less order, and more entropy. If you pour a can of black paint and a can of white paint into a bucket, you don't end up with black on one side of the bucket and white on the other side, but a grey mixture. This mixture has higher entropy. Similarly, the completely mixed up light that emerges from the pinhole in the form of cavity radiation has very high entropy. One of the most important features of the statistical approach to thermodynamics is that systems are more likely to be found in high entropy states than in low entropy states.

You have to give Planck credit for biting the bullet and using the ideas and techniques he abhorred to work out a physical basis for the equation for black-body radiation. But what he discovered was startling. According to this interpretation of events, light was not being emitted or absorbed in a continuous fashion, but in the form of little lumps, which he called 'quanta'. Speaking to the Berlin Physical Society again on 14 December 1900, Planck described his new work, and said:

> We therefore regard – and this is the most essential point of the entire calculation – energy to be composed of a very definite number of equal finite packages, making use for that purpose of a natural constant $h = 6.55 \times 10^{-27}$ erg sec.

Don't worry about the units for h; just notice how tiny it is – a decimal point followed by 26 zeroes before you get to the 6. And the 'finite packages' are only equal for each colour, or wavelength, of light. The size of each packet of energy E is given by the equation $E = h\nu$, where ν is the frequency (proportional to 1 divided by the wavelength) of that particular colour of light. Instead of establishing that matter is continuous, on the face of things Planck had found that electromagnetic radiation was not continuous! But he didn't interpret his results that way.

Looking back from 1931, Planck described his breakthrough as 'an act of desperation' and said the idea of the quantum of energy was:

> A purely formal assumption and I didn't give it much thought, except only that, under all circumstances

and at whatever cost, I had to produce a positive result.[7]

He still thought of light as a continuous wave (after all, Young's experiment still worked!), and regarded the quanta as some kind of mathematical tool, only useful in the statistical process of adding up the contributions of all the harmonic oscillators. Nobody else knew what else the quanta could actually be either; but Einstein, who had studied Kirchoff's work while he was at the ETH, heard news of Planck's work while he was teaching at Winterthur in 1901, and was deeply puzzled. He kept the problem of what Planck's quanta really were in his mind all the time he was developing his own statistical skills with the work described in his first scientific papers. Since nobody else had made any progress with interpreting Planck's equation (indeed, physicists were largely too puzzled even to try to explain it), when he had learned enough to be able to tackle the puzzle properly at the beginning of 1905 Einstein was able to pick up the thread exactly where Planck had left off.

The first great paper Einstein wrote in 1905 (he finished it on 17 March) is often referred to as the 'photoelectric paper', not least because when Einstein eventually got his Nobel Prize, the citation focused on that aspect of his work. But the section on the photoelectric effect was only a relatively small part of the paper: one of several examples that Einstein used to illustrate the importance of the concept of light quanta. Nevertheless, it was a very important part of the paper, and these ideas had also been turning over in his mind since 1901.

It all started, as I hinted earlier, with the work of Philip Lenard, a German physicist who carried out a series of experiments, starting in 1899, which investigated the way ultraviolet light shining onto the surface of a metal in a vacuum could cause it to emit electrons (then still known as cathode rays; remember that electrons had only been identified as particles in 1897!). The immediate impact Lenard's work (of which more shortly) had on Einstein when he learned about it in 1901 can be seen from the beginning of a letter he wrote to Mileva late in May that year:

> I just read a wonderful paper by Lenard on the gen-
> eration of cathode rays by ultraviolet light. Under
> the influence of this beautiful piece I am filled with
> such happiness and joy that I must absolutely share
> some of it with you.[8]

What's especially interesting about that letter is that it is the first one he wrote to Mileva after she informed him that she was pregnant; he gets around to offering his response to that news only later on in the letter. It is quite clear where Einstein's priorities lay. But the important point in terms of the genesis of the 'very revolutionary' paper on the light quantum is that this was not something that sprang suddenly into his mind fully formed at the beginning of 1905, but was a project that he had been working on, from time to time, for four years. Einstein was a genius, but he was a methodical and hard-working genius.

In some ways, he was also cautious, as we have seen with his choice of title for the Brownian motion paper. Although he knew his work on the light quantum was revolutionary, he was careful to choose a title for the photoelectric paper that would not be an immediate turnoff to readers of the *Annalen der Physik*. He settled on 'On a Heuristic Point of View Concerning the Production and Transformation of Light', thereby giving the impression that all he was offering was a convenient mathematical device for carrying out calculations, not a suggestion that light might really be made up of a stream of particles. But he immediately pulled the rug from under that cosy assumption in the opening paragraphs of the paper.

First, he spelled out the dilemma confronting physics: 'While we consider the state of a body to be completely determined by the positions and velocities of an indeed very large yet finite number of atoms and electrons, we make use of continuous spatial functions to determine the electromagnetic state of a volume of space.' Which means that this 'electromagnetic state' can never be described by a finite number of quantities, no matter how big that number is. Einstein spelled out that the energy of a material body cannot be broken down into an arbitrarily large number of arbitrarily small parts, but that in contrast – according to Maxwell's equations – the energy of light spreading out from a source gets weaker and weaker indefinitely as it spreads. These two concepts of energy had to come into conflict where matter and light interact with one another, as in the interior of a 'black-body' cavity, or in the photoelectric effect.

Then Einstein made a key point. Of course, he acknowledged, every traditional optical experiment (such as Young's experiment) produced results that were consistent with the wave model of light. But: 'One should keep in mind, however, that optical observations refer to time averages rather than instantaneous values.' In other words, light could indeed be composed of tiny particles, provided those particles were so small that their combined effects averaged out on the scale of the usual optical experiments to give the appearance of a smooth continuum. Einstein then gave a list of several recent experiments, including studies of black-body radiation and Lenard's work on the photoelectric effect, which conflicted with the classical view of light as a wave, and warned his readers what was coming:

> According to the assumption considered here, in the propagation of a light ray from a point source, the energy is not distributed continuously over ever-increasing volumes of space, but consists of a finite number of energy quanta localized at points

of space that move without dividing, and can be absorbed or generated only as complete units.

In other words, if you had infinitely sensitive eyes and looked at a source of light from very far away, you would not see a continuous faint glow, but individual flashes of light, with total darkness in between, as individual light quanta arrived at your eyes.*

If that didn't whet the appetite of his potential readers, nothing would. Without more ado, Einstein plunged into the mathematical part of his paper. He began by taking a fresh look at Planck's calculations, correcting some errors in Planck's own work and coming up with a new derivation of the key equation, describing the black-body curve, in which Planck's constant appears. Then, in a tour de force argument at the heart of his paper (the bit for which he should have got the Nobel Prize), Einstein compares the entropy of a certain volume of monochromatic radiation (a box full of light of a single colour) with the entropy of a certain volume of gas. He calculates the entropy of the radiation for short-wavelength radiation (in the region where Wien's law applies) and the entropy of an equivalent box of gas from, what were by then, the

---

* Incidentally, this is exactly what astronomers do now 'see', using sensitive detectors called charge-coupled devices, when they point their telescopes towards the faintest and most distant objects in the Universe. They can literally count the photons arriving one by one.

standard techniques developed by Boltzmann – and he gets the same answer. His conclusion is that:

> Monochromatic radiation of low density (within the range of validity of Wien's radiation formula) behaves thermodynamically as if it consisted of mutually independent energy quanta.

This is arguably the most revolutionary sentence written in science in the 20th century, given the success of Maxwell's equations. What Einstein is saying is that, as far as thermodynamic properties such as entropy are concerned, a gas behaves as if it is made up of very many tiny particles (atoms and molecules), and electromagnetic radiation behaves as if it is made up of very many tiny particles ('atoms of light,' or photons). At this basic level, there is no fundamental difference between matter and light after all. And there is another point – the most important point of all – which might easily be missed by the casual reader. Einstein has reached this conclusion without having to assume anything at all about the way light interacts with the 'harmonic oscillators' that are at the heart of Planck's treatment of the problem. He is explicitly saying that this graininess of light is an intrinsic property of the light itself, not something to do with the way light interacts with matter. He hasn't just reproduced Planck's result, but has done something much more fundamental than Planck ever did.

It is only after dropping this bombshell that Einstein goes on to consider the implications of his discovery for other areas of physics, most notably the photoelectric effect. He did not, as many people imagine, come up with the light quantum idea from the photoelectric effect, but used the photoelectric effect to demonstrate the power of this idea. And it still perfectly demonstrates the power of this 'heuristic point of view' concerning the nature of light.

The curious thing that Lenard had discovered, which had made Einstein so excited in 1901, was that the energy of an electron produced by the photoelectric effect does not depend on the intensity of the light (how bright it is), but it does depend on the wavelength of the light. The effect only happens at all for ultraviolet light, which covers a range of wavelengths even shorter than the wavelength of blue light and cannot be seen by the human eye; so 'colour' is not really the right word to use, but in a sense the energy of the electrons produced by the photoelectric effect depends on the colour of the light shining on the metal surface. Since the energy of the electrons determines their speed, you can also say that the speed with which the electrons are ejected depends only on the colour of the light shining on the surface.

Lenard's discovery runs counter to common sense, because a bright light carries more energy than a dim light. You would expect a bright light to knock electrons out of the metal surface more energetically, so they would move away faster. But Lenard found no such effect. If he used ultraviolet light with a particular wavelength, the ejected electrons always escaped with the same speed. If he turned the brightness of the light up, more electrons were ejected, but still with the same speed; if he turned the brightness of the light down, fewer electrons were ejected, but still with the same speed.

This was impossible to explain on the wave model of light. But it was utterly simple to explain using Einstein's 'heuristic principle'. Planck's equation implied that electromagnetic radiation only existed in little packets of energy, quanta with $E = h\nu$. For a particular wavelength (or frequency) of radiation, every quantum has the same energy. So energy could be handed over to electrons only in quanta that size. As Einstein put it:

> The simplest conception is that a light quantum transfers its entire energy to a single electron.

In which case, as long as ν was the same, every electron would receive the same amount of energy and would rush away from the metal surface with the same velocity. If you turned the brightness of the light up, there would be more quanta, but they would still each have the same energy $h\nu$, so there would be more ejected electrons, but each still moving with the same velocity. The only way to make the electrons move faster would be to use a different wavelength of light, with a bigger value of the frequency ν (which means a shorter wavelength, further into the ultraviolet).

This was very nearly what Lenard had found, but not quite. The trouble was, the experiments involved were still extremely difficult, and although Lenard had found that shorter-wavelength ultraviolet light did produce ejected electrons with more energy, he couldn't measure the energy precisely. The results weren't good enough to say exactly how much extra energy the electrons got for a particular change in wavelength. Einstein's calculations did predict a very precise relationship, but it was more precise than the experimental results. His equation agreed with the experimental data, but within the range of uncertainty allowed by the data it was conceivable (if unlikely) that some other equation might work just as well. So all he could say in 1905 was that:

> As far as I can tell, this conception of the photo-electric effect does not contradict its properties as observed by Mr Lenard.

With no absolute experimental confirmation of the validity of Einstein's calculations, the idea of light quanta (especially 'particles' of light that still somehow had wavelengths associated with them) was so shocking to physicists in 1905 that Einstein's paper was largely ignored for years. The only person who really took much notice of it was an American experimental physicist, Robert

Millikan, who was so infuriated when he heard about it that he promptly set out to try to prove Einstein was wrong. But it's an indication of how little attention was paid to Einstein's photoelectric paper when it was published that Millikan, who was already working on the photoelectric effect and investigating other properties of electrons in 1905, didn't even learn about the paper for several years.

Millikan worked at the University of Chicago and was 37 in 1905, eleven years older than Einstein and ten years younger than Planck. But it was only in 1912, when he was 44, that he began his determined effort to measure the properties of electrons ejected by the photoelectric effect accurately enough to test Einstein's predictions and – he firmly expected – to prove Einstein was wrong. In a classic example of the scientific method at work, the sceptical Millikan actually found that the relationship between the energy of the ejected electrons and the wavelength of the radiation involved exactly matched Einstein's predictions. But even then, he could not at first accept the reasoning behind Einstein's prediction. When he announced the results of four years of intensive research into the problem in 1916, he said:

> The Einstein equation accurately represents the energy of electron emission under irradiation with light [but] the physical theory upon which the equation is based [is] totally unreasonable.

Nevertheless, he admitted that his results, combined with Einstein's equation, provided 'the most direct and most striking evidence so far obtained for the reality of Planck's h'.[9]

The Nobel Committee were no less cautious when they awarded Einstein the Physics Prize in 1922 (it was actually the 1921 Prize, held over for a year; Millikan received the Prize in 1923). The citation noted the work of Millikan in proving Einstein's prediction right, but referred only to 'the discovery of the law of the photoelectric effect' (in other words, the equation tested by Millikan) and avoided mentioning the physical model on which the equation was based. But as it happened, just a year later, in 1923, new experiments involving the interaction between electromagnetic radiation and electrons finally established the reality of light quanta, which were then given the name 'photons' by the American chemist Gilbert Lewis in 1926.

Although it is not strictly relevant to our story, it's amusing to see how Millikan rewrote his own history of these events as the reality of photons became more and more firmly established. Having entirely dismissed the physical theory on which the equation was based in 1916, in 1949 he wrote in an article in the journal *Reviews of Modern Physics* that: 'I spent ten years of my life testing that 1905 equation of Einstein's and contrary to all my expectations, I was compelled in 1915 to assert its unambiguous verification in spite of its unreasonableness.'

In 1951, two years before he died, he wrote in his autobiography that: 'I think it is correct to say that the Einstein view of light quanta, shooting through space in the form of localised light pulses, or, as we now call them, photons, had practically no convinced adherents prior to about 1915, by which time convincing experimental proof had been found.' No mention here that even Millikan himself had still not been convinced of the reality of light quanta in 1915!

As I have mentioned, part of the problem of convincing scientists that light quanta were real was the enormous success of the wave model of light, and in particular Maxwell's equations. At the end of the 19th century, it seemed quite clear that particles were particles and waves were waves. When cathode rays were discovered, nobody knew if they were waves or particles until J.J. Thomson devised the experiments which proved that they were particles. Then, they could be neatly labelled and the possibility that they might be waves forgotten. It was equally natural to assume that light could only be one thing or the other. It wasn't until well into the 1920s that physicists began to come to terms with the uncomfortable truth that it was possible for light to somehow be both a wave and a particle, and to realise that the everyday laws of common sense do not apply on the very small scale of entities such as photons and electrons. As we shall see, Einstein also played a key part in these discoveries.

This 'wave-particle duality' lies at the heart of

quantum physics, and it is now well established that just as light (which was formerly thought of as a wave) behaves like a stream of particles under some circumstances, so electrons (and other entities that were formerly thought of as particles) have a dual nature and behave under some circumstances like waves. Electrons can even be made to interfere with one another in a variation on Young's experiment.* Einstein was the first person to understand that light could behave as a wave under some circumstances (as in Young's experiment) and like a particle in other circumstances (as in the photoelectric effect). This flexibility of approach allowed him to keep faith with the aspects of the wave model that worked – notably Maxwell's equations – even while he was rejecting the wave model in situations where it did not apply. With the photoelectric paper submitted for publication in mid-March 1905, Einstein's long fascination with light was about to bear fruit in an even more spectacular way – once he had finished writing up the paper that would become his PhD thesis and his paper on Brownian motion.

The special one

The last of the four great papers of Einstein's *annus mirabilis* emerged from his fertile brain soon after he

---

* Deliciously, while J.J. Thomson received the Nobel Prize for 'proving' that electrons are particles, his son George received the Nobel Prize for 'proving' that electrons are waves. They were both right.

had submitted his paper on Brownian motion to the *Annalen der Physik*. The breakthrough was triggered, he later recalled, by a discussion with his old friend Michele Besso, sometime in the middle of May.* An intense burst of activity over the next six weeks saw the key paper on the Special Theory of Relativity delivered to the *Annalen der Physik* on 30 June (after Mileva had carefully checked Einstein's calculations for slips, a mundane task which didn't even earn her an acknowledgement in the paper). It was published at the end of September, in the same week that the editor of the *Annalen* received a second paper on the subject, in which Einstein spelled out the famous relationship between mass and energy.[†]

Curiously to modern eyes, this key paper about the nature of space and time is actually titled 'On the Electrodynamics of Moving Bodies'. This reflects the importance of light – an electromagnetic entity – in

---

* The genesis of the Special Theory was described by Einstein in a lecture in Japan, in 1922; the lecture was reprinted in *Physics Today* in August 1982.
† Of course, the 'Special Theory' paper was not known by that name at the time; Einstein introduced the name in 1915, to distinguish it from his General Theory of Relativity. But I will use the name, since, as with the 'Brownian motion' paper, we have the benefit of hindsight. I emphasise that 'Special' in this context means the theory is a 'special case' dealing only with objects travelling at constant velocities; the General Theory deals with accelerations as well. But I reiterate that it is always 'Special Theory of Relativity;' there is no such thing as the 'theory of special relativity,' since it is the theory that is 'special' not the relativity!

Einstein's theory, but also highlights the way in which the puzzle of relative motion had developed in the 1890s. Following the success of Maxwell's equations, physicists in the last quarter of the 19th century were convinced that light was a form of vibration in the ether, and various experiments were carried out to try to measure the motion of the Earth through the ether. If light travels at a certain fixed speed through the ether (as Maxwell's equations seemed to imply), and the Earth is moving in the same direction, then you would expect from everyday experience of how speeds add up that the speed of that light relative to the Earth would be less than the speed of that light through the ether. Conversely, if the Earth were running head on into a light beam travelling through the ether, then you would expect the speed of the light beam measured in the experiments to be equal to its speed through the ether plus the speed of the Earth through the ether.

Making such measurements proved extremely difficult (chiefly because the speed of light is so big, 300,000 kilometres per second), but the predictions encouraged experimenters to develop new techniques in the 1880s and 1890s which were accurate enough to measure the calculated effects. But even when these experiments became sophisticated enough to take account of things like the Earth's movement around the Sun, and its daily rotation on its own axis, they always measured the same velocity for light, whether it was moving in the same direction as the Earth, in the opposite direction to the Earth, or at any angle across the line of the Earth's motion.

The first person to take these results seriously and try to find an explanation for what was going on (rather than just assuming the experimenters were making a mistake) was George Fitzgerald, who was professor of Natural and Experimental Philosophy at Trinity College, Dublin. In 1889 he wrote a paper, which he sent to the American journal *Science*, in which he pointed out that the experimental results could be explained if the experimental apparatus (and everything else) shrank slightly in the direction of its motion through the ether. The experimental apparatus involved is much more complicated than a simple ruler, but in effect he said that if your ruler shrank by a tiny amount then the time taken for light to whizz past the ruler from one end to the other would be a little less, and you would measure a different speed than if the ruler had not shrunk. 'Paper', though, is perhaps too grand a word for Fitzgerald's squib, which contained no mathematical calculations and consisted only of a single paragraph. It was clear that in order for all experiments to always measure the same speed for light, the shrinking had to obey a precise mathematical formula, which Fitzgerald did not spell out. This shows, for example, that in order to make a metre-long ruler shrink to 99 centimetres (that is, a reduction in length of just 1 per cent), it would have to be moving at one seventh of the speed of light, 43,000 kilometres per second.

Fitzgerald had no detailed physical explanation for why objects should shrink in this way, and his colleagues in Dublin laughed at the idea. Although the paper was published in *Science*, nobody took any notice.* So when the Dutch physicist Hendrik Lorentz came up with a similar idea in 1892, and the appropriate mathematical equation to describe the shrinking effect, he didn't know about the similarity to Fitzgerald's earlier work until this was pointed out by the British physicist Oliver Lodge.

The contraction formula became known as the Lorentz-Fitzgerald contraction, which seems a little unfair both in terms of the chronology and alphabetically. There was, though, a sound reason why it became known as Lorentz-Fitzgerald contraction, not Fitzgerald-Lorentz contraction. Unlike Fitzgerald, Lorentz developed this mathematical equation alongside a physical picture of what might be going on to make moving objects shrink.

In the early 1890s, Lorentz, who was born in 1853, was professor of Theoretical Physics at the University of Leiden. He had developed a theory of electrodynamics which he was already calling the 'electron theory' – although, confusingly from our point of view, this did not involve the particles now known as electrons. He

---

* Indeed, for some time after his paper had been published on the other side of the Atlantic, Fitzgerald himself didn't know that it had appeared in print.

suggested that all matter is made up of electrically charged particles, some with positive charge and some with negative charge, held together by electromagnetic forces. In 1892, he simply referred to these entities as 'charged particles', but in 1895 he referred to them as 'ions'.* It was only in 1899, two years after the identification of cathode rays as streams of negatively charged particles, that he started calling these particles 'electrons'. The name stuck, even though the 'electron theory' did not last.

Like Fitzgerald, Lorentz argued that the experimental observations of the constancy of the speed of light could be explained if moving objects shrank in the direction they were moving, and he came up with the appropriate formula for the contraction. Also like Fitzgerald, he assumed that the cause was the motion of the object relative to the ether, which provided a standard frame of reference against which all motion could, in principle, be measured. But he went further by suggesting that the reason why objects shrank in this way was because of a physical effect of the ether on the moving objects. Specifically, he suggested that there was an electric force which was caused by the motion, which had the effect of squeezing the charged particles of which the moving object was made. Lorentz took these ideas much further than Fitzgerald (not least because Fitzgerald died

---

* Again, with a different meaning from what 'ion' means to a scientist today.

in 1901, at the early age of 49) and came up with a complete theory that he published in a Dutch journal, which was not very widely read, in 1904. As well as describing how the length of a moving object was related to its motion relative to the ether, this work raised the idea of 'relative time' and the idea of synchronising clocks by using light signals – which was also, as we shall see, a central feature of Einstein's work.

In all of this, Lorentz was encouraged by the French mathematician Henri Poincaré, a year younger than Lorentz, who publicised the ideas and became interested in the mathematical foundations of the equations involved. Indeed, the first appearance of the term 'relativity principle' was in a lecture Poincaré gave at the World Exhibition in St Louis in 1904. It was also Poincaré who first used the term 'Lorentz transformations' to describe the whole package of equations that Lorentz had derived in his 1904 paper.

Einstein had followed at least some of these developments, and was well aware of the fact that all experiments showed no effect of the motion of the Earth on the measured speed of light, even though he does not seem to have been particularly familiar with the details of all the experiments. He later said that he had spent seven years puzzling over the electrodynamics of moving bodies before the breakthrough in 1905, and this is borne out by a letter he wrote to Mileva in August 1899, in which he said:

I'm more and more convinced that the electrodynamics of moving bodies as it is presented today
doesn't correspond to reality, and that it will be possible to present it in a simpler way. The introduction
of the term 'ether' into theories of electricity has led
to the conception of a medium whose motion can
be described without, I believe, being able to ascribe
physical meaning to it.[10]

This is where Einstein would make his dramatic breakthrough in 1905. He did away with the ether. Instead of
saying that what matters is motion relative to the ether, he
said that what matters is how two objects move relative to
each other, and that there is no absolute standard of rest
against which motion can be measured.

There's another especially intriguing feature of the
paper on what became known as the Special Theory. It
contains no references at all to any earlier work, not even
that of Lorentz, or the experiments involving the speed of
light. Instead, it starts from first principles and Maxwell's
equations to build a logical, consistent mathematical
structure which leads inevitably to the conclusions about
the nature of space and time. By structuring his paper in
this way, Einstein is clearly proclaiming to the world that
he has discovered a fundamental, absolute truth about the
nature of the Universe, to rank with such fundamental
mathematical truths as Pythagoras' theorem concerning
the lengths of the sides of right-angled triangles. It does

not depend on experiments or theoretical models, it is part of the very fabric of the way the world works.*

In this spirit, Einstein starts with just two facts about the world – what the mathematicians would call postulates or axioms – and constructs the whole edifice of his theory by building upward from those foundations. The first postulate comes straight from the world of the practical application of electromagnetism in dynamos and electric motors, the industry where Einstein's father had worked for so long. The 19th-century boom in this industry was based on the work of Michael Faraday, who discovered in 1831 that when a conducting wire moves in a magnetic field, an electric current flows in the wire. Or rather, using modern terminology, he found that when a wire moves relative to a magnetic field an electric current flows in the wire. It doesn't matter whether the magnet is fixed in place in the laboratory and the wire moves past it, or whether the wire is fixed in place in the laboratory and the magnet moves past it. Either way, an electric current flows in the wire. As Einstein put it in the opening paragraph of his paper on the Special Theory:

---

* Shortly before he died, Einstein told his biographer Carl Seelig that in 1905 he knew about Lorentz's work of 1895, but not about his later work or Poincaré's contributions. This may be an exaggeration, but he probably had not actually read Lorentz's 1904 paper, since the *Proceedings* of the Amsterdam Academy were not exactly easy to get hold of in Bern.

The observable phenomenon here depends only on
the relative motion of conductor and magnet.

And this leaves no role for the ether, since the observed
phenomenon is not affected by the motion of either the
magnet or the wire relative to the ether. If, for example,
both the wire and the magnet move alongside each other
in the same direction (any direction!) and at the same
speed (any speed!), there is no current in the wire.

This brings us to the idea of reference frames. A reference frame is the place you make measurements from,
like the laboratory in the previous example. We know that
the lab is actually being carried along with the Earth's
motion, but we can treat it as if it were at rest. All the
laws of mechanics (Newton's laws) work perfectly in the
lab. Those same laws also work perfectly in a frame of
reference moving at a constant velocity relative to the lab
– for example, in an aircraft flying at a smooth and steady
speed, or, in the example Einstein always favoured, in a
train rolling smoothly along a track at a constant speed in
a straight line (that is, at constant velocity). Such frames,
in which Newton's laws work perfectly, are now called
'inertial frames'.

The first person to spell most this out, long before
the days of trains and planes, was Galileo Galilei, in the
17th century; his insights were part of the foundations of
Newton's work. As far as the physical behaviour of matter and Newton's laws are concerned, by 1905 it had been

known for hundreds of years that there is no difference in how things behave if your frame of reference is a lab in an immovable building or another lab in another frame of reference moving steadily at a constant velocity relative to the first lab. Newton himself believed that there must be an absolute 'standard of rest' in the Universe against which all motion could be measured, and this tied in with the idea of the ether. But no experiment involving Newton's laws could ever detect motion relative to this hypothetical absolute rest frame.

Now, in 1905, Einstein had realised that no experiment involving Maxwell's equations could ever detect motion relative to this hypothetical absolute rest frame, either. So there was no need for the ether. Referring to the way electricity is generated by the relative motion of a wire and a magnet, he continued:

> Examples of this sort, together with the unsuccessful attempts to detect a motion of the earth relative to the 'light medium', lead to the conjecture that not only the phenomena of mechanics but also those of electrodynamics have no properties that correspond to the concept of absolute rest. Rather, the same laws of mechanics and optics will be valid for all coordinate systems in which the equations of mechanics hold.

In this context, 'coordinate system' means the same as

'reference frame'. Einstein is saying that all of the laws of physics, both for mechanics and electrodynamics, are the same in any reference frame moving at constant velocity relative to any other reference frame in which those laws apply. There is no special reference frame which can be regarded as at rest in an absolute sense. He called this his first postulate, which, echoing Poincaré, he named the 'principle of relativity'.

The second postulate is even simpler, and comes straight from Maxwell's equations:

> Light always propagates in empty space with a defi-
> nite velocity V that is independent of the state of
> motion of the emitting body.

Einstein based this postulate on Maxwell's equations, not on the experiments that tried to detect the effect of the Earth's motion on the measured speed of light. As he realised, all that could be said about those experiments was that they had 'failed to detect' any movement of the Earth relative to the hypothetical ether, and failed to measure any change in the speed of light. But there could still be effects too small for the experimenters to have measured. The postulate, though, was precise and unequivocal – as were the results he obtained from these two disarmingly simple assumptions.

The other key ingredient of the Special Theory is a direct consequence of the second postulate. For objects moving at constant velocity relative to one another, each object can be regarded as carrying its own reference frame (coordinate system) along with it. Einstein's theory had to deal with the relationship between coordinate systems, clocks and electromagnetic processes. And the first thing he had to do to develop this theory was to spell out the way the second postulate affects our ideas about time.

In everyday life, we all have an idea in our heads that it is the same 'time' everywhere at once. But what does this really mean? If some master clock in London sends out a time signal by radio at noon, and I have a clock which receives that radio signal and automatically sets itself to noon, my clock will actually be a fraction of a second behind the clock in London, because it takes radio waves, travelling at the speed of light, a certain amount of time to reach my clock. As long as I don't move my clock, and I know how far it is to London, I can get round this problem by building in an allowance for the time taken by the radio waves to travel to the clock. Since light travels at a constant speed, the only completely accurate way to do this is to send a light signal to London and back, time how long the return journey takes and divide by two to work out the difference. But what if this process is watched by a moving observer?

The first postulate says that this observer is entitled to regard himself as stationary, with the clocks moving past him at a constant velocity. He will see the calibrating light pulse from the clock head off to London and bounce back, but the return journey will be a different distance in his frame of reference because the whole moving coordinate system will have shifted forward while the light was on its journey. So the 'stationary' observer and the 'moving' observer do not agree on the distance between the clock and London, and they do not agree on what 'the same time' means. In other words, measurements of both time and space are relative – they depend on how the person who makes them (the observer) is moving, relative to the things they are measuring. Remember, also, that the second postulate says that each observer will see all light pulses, whether they originate in his own frame of reference or the other one, moving at the same speed. Putting the appropriate mathematics into all of this led Einstein to find a system of equations that could be used to transform the measurements made by one observer into the equivalent measurements in any other coordinate frame moving at constant velocity relative to that observer.* These coordinate transformations are exactly the same as the equations found by Lorentz a year earlier, although Einstein had not read Lorentz's 1904 paper in 1905. But there is a huge difference in the interpretation of those equations.

Lorentz had found a system of equations that worked,

---

* I do not have space to go through the argument in detail here; the easiest way to understand what is going on is provided by Lewis Epstein in his book *Relativity Visualized*.

but which were solely based on the need to explain the experimental failure to measure the motion of the Earth relative to the ether. But he still thought in terms of the ether, and there was no underlying principle shoring up his equations. This is a little like the way Planck came up with an equation for the black-body curve that fitted the curve but had no foundation in terms of an underlying principle. By contrast, as he had with the black-body radiation, Einstein started out from first principles and proved that the world really must work in accordance with the equations.

The best example of how different Einstein's view of the world was even to that of Lorentz and Poincaré is that he understood the importance of the symmetry in what we still call the Lorentz transformations. Poincaré had noticed that the transformations are symmetrical. If one observer sees a moving object shrunk in the direction of its motion, then the symmetry implies that an observer riding with the moving object sees his own frame of reference as perfectly normal, but the first observer and everything else in his coordinate system shrunk. Poincaré dismissed this as a quirk of the equations, with no physical significance – it made no sense to him to turn his world view around and think of the ether being 'shrunk' as it moved past us. But Einstein, who had no need of the ether, saw the symmetry in the transformations as a fundamental truth of profound physical significance.

Einstein proved that an observer in one inertial frame would perceive objects in a different inertial frame shrunk in the direction of their motion relative to him, and he would see clocks in the other inertial frame running more slowly than clocks in his own inertial frame. An observer in the other inertial frame would see the mirror image of this – he would see the first observer's clocks running slow, and the first observer's rulers and other equipment (and, indeed, the first observer) shrunk. All of this has now been confirmed by experiments. The effects are very small unless the relative velocities involved are a sizeable fraction of the speed of light (which is why we don't notice them in everyday life and they are not common sense), but particles are routinely observed travelling at such speeds in 'atom smashing' machines like those at CERN, in Geneva, and Fermilab, in Chicago. The Special Theory has been proved accurate time and time again. It also makes one further prediction, which Einstein himself didn't spot in June 1905, but which he quickly realised and wrote about in another paper, which was a kind of footnote to the paper on the Special Theory. That realization led to the most famous equation in science – although, disappointingly, the equation does not appear in its familiar form in the paper itself.

In the summer of 1905, Einstein wrote a letter to Conrad Habicht in which he said:

One more consequence of the paper on electro-dynamics has also occurred to me. The principle of relativity, in conjunction with Maxwell's equations, requires that mass be a direct measure of the energy contained in a body; light carries mass with it. A noticeable decrease in mass should occur in the case of radium. The argument is amusing and seductive; but for all I know the Lord might be laughing over it and leading me around by the nose.[11]

Einstein mentions radium because this archetypal radio-active element emits energy in the form of radiation and heat all the time. The origin of this energy had been a puzzle for science ever since the discovery of radium by Marie and Pierre Curie at the end of the 1890s; Einstein's discovery implied that the matter radium was made of was slowly being converted into energy, so that the radium itself would gradually lose mass. There had previously been suggestions that electromagnetic energy might be associated with mass, and even that the electron's mass might be completely attributed to its electromagnetic field. But Einstein was suggesting something different: that all matter had an energy equivalent, an energy that might in principle, be liberated; and he calculated a precise value for this energy.

His little paper pointing this out (just three pages long in its printed form) was received by the *Annalen der Physik* on 27 September. It was published before the end of the year, but in the next volume of the *Annalen* (volume 18) from the volume containing the three great papers on

Brownian motion, light quanta and the Special Theory (volume 17, now a valuable collectors' item). Still using V to denote the velocity of light, in his own words Einstein concluded:

> If a body emits the energy L in the form of radiation, its mass decreases by L/V^2. Here it is obviously inessential that the energy taken from the body turns into radiant energy, so we are led to the more general conclusion:
>
> The mass of a body is a measure of its energy content; if the energy changes by L, the mass changes in the same sense by $L/9 \times 10^{20}$ if the energy is measured in ergs and the mass in grams.

In those units, the speed of light is 3×10^{10} cm per second, and 9×10^{20} is the speed of light squared. Putting E for the energy rather than L, c for the speed of light rather than V, m for mass, and re-arranging the equation slightly to have the energy, rather than the mass, on the left hand side, the equation Einstein discovered becomes:

$$E = mc^2$$

The only equation that everybody knows.

Like all of the results from the papers written by Einstein in his *annus mirabilis*, this prediction has since been amply confirmed by experiment – not least the awesome 'experiment' of the nuclear bomb. It is now understood that the conversion of mass into energy provides the energy source which keeps the Sun and stars shining, and is therefore the ultimate source of the energy on which life on Earth depends. Which makes this little paper just about the most important 'footnote' in scientific history.

As I hope I have made clear, however, all of the work that Einstein produced in 1905 was of its time. The statistical ideas which underpinned the doctoral thesis and the work on Brownian motion were a significant step forward, but still part of the mainstream of the investigation of atoms and molecules. The light quantum paper jumped off from the work of Max Planck, and also used statistical ideas from thermodynamics. Even the paper on the Special Theory – although strikingly different in its foundations from the approach used by Lorentz and Poincaré – came up with the same transformation equations, and it is not unlikely that something similar to the Special Theory would soon have emerged from the line of thinking pioneered by Poincaré himself. In isolation, each contribution was something that an individual physicist at the height of his powers might have been proud of, as the biggest achievement of his career – even the relatively mundane PhD paper, since to many physicists a PhD is the greatest achievement of their academic lives.

What made Einstein so special, and the *annus mirabilis* so miraculous, was that all four pieces of work were produced by the same young man, working out-side the mainstream of scientific life, in his spare time, while holding down a demanding job at the patent office which required his attendance there six days a week for eight hours a day. Not to mention the inevitable distur-bance caused by a year-old baby back at his apartment. And it wasn't as if it really took him a year to do all this. Although, admittedly, a lot of prior thought had gone in to all this work, the four great papers were actually written between March and June 1905, at a rate of one a month, and the $E = mc^2$ paper was finished just three months later, at the end of September.

It would take a little while for the importance of all this to sink in, and for the true genius of Einstein to be appre-ciated. Indeed, he even stayed at the patent office (partly by choice) for another four years before at last becoming a university professor at the age of 30. He would also continue to make major contributions to physics until he was in his forties, a remarkably long time for any theoreti-cal physicist. The rest of his life, and his place in history, would be forever coloured by that outburst of creativity in 1905; but it would be ten years before, in 1915, he pro-duced something really special (in the everyday sense of the term!): a theory that was very much not of its time, and which in all probability would not otherwise have emerged for decades, if at all in the form that we know it. Even for Einstein, though, the road to the General Theory was, as we shall see, far from straightforward.

3 The Long and Winding Road

The geometry of relativity; Moving on; In the shadow of a giant; On the move; First steps; What Einstein should have known; The masterwork

The work Einstein published in his *annus mirabilis* didn't immediately set the scientific world on fire, but it was noticed and drew him into correspondence with a widening circle of physicists, many of whom were astonished to find that he was a junior patent officer and not a professor at the University of Bern. In May 1906, he wrote in a letter to Lenard that: 'My papers are meeting with much acknowledgement and are giving rise to further investigations. Prof. Planck [Berlin] wrote me about it recently.'[1] The paper Planck was particularly interested in was not, however, the one on light quanta, about which he had reservations, but the Special Theory of Relativity, of which he was an early and enthusiastic champion.

Einstein himself continued to publish, though not at the prodigious and unsustainable rate of 1905, and his first paper really to make waves in the scientific community appeared in 1907. In this work, Einstein used the idea of energy quanta, combined with his now familiar statistical approach, to explain the way the temperature of an object increases as it absorbs heat. The kinetic theory explains in a qualitative way that the rise in temperature of a solid body as it absorbs energy means that the atoms and molecules vibrate more strongly when the material is hotter. Einstein introduced the idea that the energy being absorbed by the individual atoms and molecules can only be accepted in quanta with energy hv. He was able to explain otherwise puzzling features of the process and found a formula for the specific heat of a body, which is a measure of how much its temperature rises when a certain amount of heat is absorbed. Planck's colleague in Berlin, the professor of Physical Chemistry Walther Nernst, took up the idea and incorporated it into his own work on thermodynamics and specific heat, which established over the next few years that it was essential to incorporate quantum ideas into any satisfactory understanding of the thermodynamic behaviour of solid objects.

The geometry of relativity

But the most important outside contribution, not only to popularising Einstein's ideas but (eventually) to shaping the way Einstein's own work would develop, came in September 1908, from Hermann Minkowski, formerly a professor of mathematics at the ETH who had been one of Einstein's teachers, but was now based at the University of Göttingen. Minkowski had been fascinated by the Special Theory as soon as he saw Einstein's paper,

1. Einstein's mother,
Pauline Einstein
*(Hebrew University of Jerusalem Albert
Einstein Archives, courtesy AIP Emilio
Segrè Visual Archives)*

2. Einstein's father,
Hermann Einstein
*(Hebrew University of Jerusalem Albert
Einstein Archives, courtesy AIP Emilio
Segrè Visual Archives)*

3. Albert as a boy, circa 1893
*(Hebrew University of Jerusalem
Albert Einstein Archives, courtesy
AIP Emilio Segrè Visual Archives)*

4. Albert and Maja Einstein
*(Hebrew University of Jerusalem
Albert Einstein Archives, courtesy
AIP Emilio Segrè Visual Archives)*

5. (*L–R*) Marcel Grossmann, Albert Einstein, Gustav Geissler,
and Eugen Grossmann in Thalwil, near Zurich
(Hebrew University of Jerusalem Albert Einstein Archives, courtesy AIP Emilio Segrè Visual Archives)

6. The house in which
Einstein lived in Bern,
49 Kramgasse *(AIP Emilio
Segrè Visual Archives)*

7. Albert with Mileva
and Hans Albert *(Hebrew
University of Jerusalem Albert
Einstein Archives, courtesy
AIP Emilio Segrè Visual Archives)*

8. Albert, circa 1912
(ETH-Bibliothek Zurich, Image Archive/J.F. Langhans)

9. First Solvay Congress, Brussels, 1911
(*L–R seated at table*) Nernst, Brillouin, Solvay, Lorentz,
Warburg, Perrin, Wien, Curie, Poincaré
(*L–R standing*) Goldschmidt, Planck, Rubens, Sommerfeld, Lindemann,
de Broglie, Knudsen, Hasenöhrl, Hostelet, Herzen, Jeans, Rutherford,
Kamerlingh Onnes, Einstein, Langevin
(Benjamin Couprie, Institut International de Physique Solvay, courtesy AIP Emilio Segrè Visual Archives)

10. Albert and Elsa Einstein aboard the SS *Rotterdam* en route to the US, 1921
(Library of Congress, courtesy AIP Emilio Segrè Visual Archives)

11. (*Front*) Eddington and Lorentz
(*Back*) Einstein, Ehrenfest and de Sitter, at the Leiden Observatory
(AIP Emilio Segrè Visual Archives, Gift of Willem de Sitter)

and astounded that a pupil he remembered as a 'lazy dog' should have come up with something so profound.[2] He accepted Einstein's ideas without question, and set about re-formulating them into a more elegant mathematical package. What he found was that everything contained in the Special Theory could be described in terms of geometry – provided that time was regarded as a fourth dimension, on a par with the three familiar dimensions of space.

The natural way to get a handle on this is to think of some everyday experiences in terms of geometrical coordinates. If you draw a triangle on a piece of graph paper, you can specify the triangle completely by giving the coordinates on the grid of the three corners of the triangle then drawing straight lines to join them up. Similarly, if you arrange to meet someone on the corner of, say, Fourth Street and Main, you are specifying a geometrical location in terms of similar coordinates in two dimensions. If you arrange to meet in the coffee shop on the second floor of the building at the corner of Fourth and Main, you are specifying the location in three dimensions, since the level in the building now comes in as an extra coordinate. And if you say you will meet in the coffee shop on the second floor of the building on the corner of Fourth and Main at three o'clock, you have introduced a fourth coordinate, the time.

Minkowski showed how the mathematics behind things like shrinking rulers and clocks that run slow could be incorporated into this kind of language, set in a framework of a four-dimensional entity, which became known as spacetime. Crucially, this means that there are properties of objects which always stay the same in spacetime, even if they look different to different inertial observers in three dimensions. A nice analogy is with the length of the shadow cast by a pencil on a wall. By twisting the pencil about in three dimensions you can make the two-dimensional shadow longer or shorter, but the pencil always stays the same length. In four dimensions, the equivalent property to length is called 'extension', and by twisting objects around in four dimensions you can change the length of the 'shadow' it casts in three dimensions. The fact that time stretches while space shrinks for a moving object reflects the fact that the four-dimensional extension in spacetime stays the same – in a sense, the two effects balance each other.

Minkowski presented his ideas in a lecture given in Cologne at the beginning of September 1908. Introducing his talk, he said:

The views of space and time which I wish to lay before you have sprung from the soil of experimental physics, and therein lies their strength. They are radical. Henceforth space by itself, and time by itself, are doomed to fade away into mere shadows, and only a kind of union of the two will preserve an independent reality.[3]

Minkowski never lived to see those words in print, since he died, from complications resulting from appendicitis, in January 1909, when he was just 44 years old. But it is no coincidence that widespread acceptance of Einstein's ideas, broad recognition of his abilities and the opening up of doors into academic life, all followed after Minkowski's re-formulation of the Special Theory.

At first, Einstein himself was slightly miffed about the way the mathematicians had picked up his ball and run off with it. Perhaps not entirely seriously, but with an undercurrent of irritation, he commented that 'since the mathematicians have attacked the relativity theory, I myself no longer understand it,' and said 'the people in Göttingen sometimes strike me not as if they wanted to help one formulate something clearly, but as if they wanted only to show us physicists how much brighter they are than we.'[4] But he soon changed his tune, when he discovered that Minkowski's geometrisation of the Special Theory was one of two important steps that would eventually put him on the road to his greatest triumph, the General Theory of Relativity. The other key step was an insight that struck him in 1907, while sitting at his desk in the patent office.

At the time, still months before Minkowski's geometrisation of the Special Theory, he was working on an article about the Special Theory for publication in an annual review of science. In a lecture he gave in Japan in 1922, Einstein said:

> I was sitting in a chair in the patent office in Bern
> when all of a sudden a thought occurred to me: 'If
> a person falls freely he will not feel his own weight.'
> I was startled. This simple insight made a deep
> impression on me. It impelled me toward a theory
> of gravitation.

It was, he said, 'the happiest thought of my life'.* But
the theory derived from that happy thought would take
Einstein almost another decade of intermittent but
often intense struggle to achieve, even with the help of
Minkowski's geometrisation of the Special Theory.

The reason why the 'sudden thought' was so important is that as soon as he had completed the Special Theory
Einstein began to attempt to find ways to make the theory
more general (hence the name) by adapting it to deal with
accelerated motion, not just motion at constant velocities. A freely-falling object is accelerating, and Einstein's
insight was the realization that this acceleration exactly
cancels out the weight of the object. Turning this around
(and remember, Einstein was working on this before the
time of space rockets, or even high-speed elevators), if you
were standing on a platform that was being accelerated
in a straight line through space, you would feel as if you
were standing still and held down by your own weight in

---

* In German, *die glücklische Gedanke*, which is traditionally quoted as 'happiest' but might be better interpreted as 'luckiest'.

a gravitational field. In 1907, Einstein realised that acceleration and gravity are exactly equivalent to one another, so that his General Theory, when he found it, would be a theory of gravity, not just a theory of motion. But the path from the insight to the theory was long and tortuous, and Einstein's personal life underwent many changes along the way.

Moving on

The first significant change came when he moved on from the patent office in 1909, four years after the *annus mirabilis*. Until then, his home life hadn't really changed much. On the strength of his doctorate, in 1906 he had been promoted to Technical Expert II Class, with an increased salary of 4,500 francs a year, but this made little difference to his frugal lifestyle. In 1908, he became a *Privatdozent* at the University of Bern – a kind of part-time, poorly paid lecturer. This was rather pointless in itself, but an essential step before he would be considered for a full university post. Curiously, though, before doing this Einstein seriously considered the idea of becoming a school teacher, rejecting the university system that had, so far, rejected him. He wrote to Marcel Grossmann that this idea stemmed from an 'ardent wish to be able to continue my private scientific work under easier conditions',[5] and actually applied for a post teaching mathematics at a Zurich high school, enclosing with his application copies of all his published scientific papers. The bemused school governors did not shortlist him for an interview, even though in his covering letter he pointed out that he would be able to teach physics as well. How different might the development of physics have been if they had had more imagination?

It was only after failing with this application that Einstein buckled down to writing the thesis that was required in order to be accepted as a *Privatdozent* at the University of Bern. This was simply an extension of his 1905 work on light quanta, and a pure formality. Einstein duly became a *Privatdozent* in February 1908, but the tiny remuneration associated with the position was not enough for him to give up the day job, so he now had more work to do and less time for science. His lectures took place on Tuesdays and Saturdays at 7am, and initially attracted an audience of three. By the summer of 1909, there was only one student attending, and the lectures were cancelled. No wonder the high school job had appealed to him.

The next step would be a professorship, which would enable him to give up the patent office job at last. Even before making that step, in July 1909 Einstein was awarded his first honorary degree (by the University of Geneva). Meanwhile, a long campaign waged by the physics professor at Zurich, Alfred Kleiner, had persuaded the university to establish a new post for a professor of theoretical physics – only a kind of junior professorship with a lower status than Kleiner's chair, but still a long way up from being a *Privatdozent*. Kleiner had one person in mind for the new post, and it was not Einstein but a young man called Friedrich Adler, who had been a student contemporary of Einstein at the ETH. But Adler, the son of the leader of the Austrian Social Democratic Party and himself politically active, decided he was better suited to political philosophy than to physics. He told Kleiner of his decision in June 1908, and at that meeting the two of them decided that Einstein was the right man for the job.

Kleiner visited Einstein in Bern to discuss the possibilities, and at the end of June attended one of Einstein's lectures. He was not impressed, and at first Einstein's lack of skill at lecturing seemed to have ruled him out. But Kleiner relented sufficiently to allow Einstein to give what amounted to an audition in the form of a lecture in Zurich. For once, Einstein prepared properly and presented the lecture adequately. 'Contrary to habit,' he wrote to a friend, 'I lectured well on that occasion.' In spite of some objections to Einstein's Jewishness, Kleiner got the appointment approved in March 1909, and the post was duly offered to Einstein – who calmly turned it down, since the salary offered was less than he was getting at the patent office. But the salary offered was increased, and he accepted. The pay was now exactly the same as that of a Technical Expert II Class at the patent office, but from taking up the appointment in the autumn of 1909 he could at last call himself Herr Professor Einstein. In the years that followed, Adler missed few opportunities to let people know that he had stood aside to make room for Einstein in Zurich.

Just before taking up the appointment, in July 1909 Einstein received his honorary doctorate from the University of Geneva, part of the celebrations of the anniversary of the founding in 1559, by John Calvin, of the Academy which evolved into the university. Einstein was greatly amused at the contrast between the teachings of Calvin and the lavish festivities laid on to mark the 350th anniversary. But though he remained as uninterested as ever in the trappings that society associated with success, he was now definitely part of the academic establishment.

A few weeks later, in September 1909, Einstein attended a conference in Salzburg, which focussed on the new developments in relativity theory and quantum physics. This gave him an opportunity to meet luminaries such as Max Planck, who he had previously known only through their publications and letters. To the surprise of the organisers, Einstein chose to provide a contribution on quantum theory, not relativity, having temporarily put aside his quest for what would become the General Theory to concentrate on what he regarded as the more pressing problem of reconciling the particle (quantum) and wave descriptions of light. The result was a lecture that became recognised as one of the landmarks of 20th-century science.

'Light,' Einstein told his audience, 'has certain basic properties that can be understood more readily from the standpoint of the Newtonian emission theory than from the standpoint of the wave theory,' and 'I thus believe that the next phase of theoretical physics will bring us a theory of light that can be interpreted as a kind of fusion of the wave and the emission [particle] theories of light.' This was the genesis of the idea of wave-particle duality. He warned that this would mean a profound change in physics, which could undermine the classical (that is, Newtonian) concept of determinism. He was right, but it took twenty years for physics to catch up with his insight, and some physicists are still arguing about the implications today. In 1909, it was all too much for Planck and many members of the audience, and Einstein, now aged 30, moved on to Zurich to take up his chair without having convinced many people. There, he turned his attention back to relativity theory, remarking to a friend: 'The more successes the quantum theory enjoys, the sillier it looks.'

Rather than settling down in Zurich and working his way up the academic ladder there, Einstein's move marked the beginning of his years as a peripatetic professor, hopping from university to university in search of security and (more importantly) the opportunity to work on what he wanted the way he wanted. It also marked the beginning of the end for his marriage; I will look briefly at the personal side before getting back to the physics.

In the shadow of a giant

Einstein's marriage had already been showing signs of strain before 1909. It wasn't easy for a woman who had once had scientific aspirations of her own, and had struggled so hard to get an education, to live in the shadow of

Einstein's growing success and fame, quite apart from the emotions associated with giving birth to and then giving away an illegitimate daughter. We now see Einstein as the iconic 20th-century scientist, an almost unique genius. But in the first decade of the 20th century, Mileva didn't even have the comfort of knowing just how special he was. Not long before, they had been equals; now, he had flown far above her. She must have thought, at least occasionally, that if it had not been for the accident of her gender she could have achieved what he had done. As Einstein's reputation grew, Mileva found herself increasingly left out of his life. When he wasn't working, he was talking about physics with his new friends, many of them young men fired with enthusiasm for their subject who must have reminded her all too painfully of her own failed ambitions in science. Things hadn't been so bad when Albert was an unknown scientist, but now he was being taken away from her. Mileva had never expected to become a housewife; there are also signs that she suffered from clinical depression, an illness even less sympathetically dealt with then than now.

Their return to Zurich, the city in which Albert and Mileva had met and fallen in love, seems to have provided a temporary revival of their relationship. Within a month of the move, Mileva was pregnant with their second son, Eduard, who would be born on 28 July 1910. But although at first the move to Zurich must have seemed like an opportunity to revisit the scene of their own youth, and perhaps make a fresh start, it didn't work out as she had hoped. The birth of Eduard was difficult for Mileva, and Eduard turned out to be a sickly baby who needed a lot of attention, while six-year-old Hans Albert could hardly be ignored.

By chance, the Einsteins had rented an apartment in the same building where Friedrich Adler and his wife lived, providing the opportunity for a firm friendship to develop between the couples, and Einstein with a sounding board on which to try out his ideas. He also renewed acquaintance with Marcel Grossmann, now a professor of mathematics at the ETH, whose notes had helped Einstein to pass his examinations. Einstein's ability as a lecturer improved a little, but what really endeared him to his students was his informality. He would allow his lectures to be interrupted by questions – almost unheard of in the German-speaking academic world at the time – and he invited any students who wanted to join him after the lectures at the Café Terasse to talk physics and set the world to rights. The birth of Eduard also brought out the fathering instinct in Einstein, after his own fashion. He was a loving – even doting – father who played with the boys and told them stories; but there are also, as we have mentioned, many accounts of occasions when he could be found rocking the crib with one hand while writing equations with the other, pushing a pram on which a notebook lay open on the blankets of the sleeping baby, or otherwise being lost in thought when he should have been paying close attention to the children.

Initially, everything had seemed fairly settled in Zurich. But in March 1910, just five months after taking up his post there and before Eduard was born, Einstein received an invitation to put himself forward for a full professorship in Prague. Things were complicated, because once again there were difficulties at home with the now-pregnant Mileva. We don't know the details, but Einstein hinted darkly about these difficulties in letters to his mother, who had by now returned to Germany and was living in Berlin. A move to Prague – a move anywhere – would surely not appeal to Mileva, but her feelings were clearly of second-ary importance, at best. The Prague post was superficially attractive as a career move – a full professorship with a larger salary in a German-speaking university (Prague was then, of course, part of the Austro-Hungarian Empire). But there were hurdles to jump. Under the notoriously bureaucratic Austro-Hungarian regime, Einstein could not simply be headhunted. The university had to pre-sent a shortlist of preferred candidates to the Ministry of Education in Vienna, which had the final say. Einstein was top of the list, with a strong recommendation from no less a physicist than Max Planck. But the ministry didn't like the idea of appointing Einstein, not least because he was Jewish. They offered the post to the second candidate on the list, a non-Jewish Austrian, Gustav Jaumann.

Meanwhile, in another twist, the students in Zurich had got wind of the possible move, and organised a peti-tion calling on the authorities 'to do your utmost to keep this outstanding researcher and teacher at our university'. The response of those authorities was to increase Einstein's salary from 4,500 Swiss francs to 5,500 SF. This was where things stood in August 1910. But Jaumann now learned that he was the second choice of the Prague faculty for the professorship and turned the offer down, declaring: 'I will have nothing to do with a university that chases after modernity and does not appreciate merit.' The job was now very nearly Einstein's – except for a final twist. Previously rejected for being Jewish, he was now on the brink of being rejected for not being Jewish enough. The Austro-Hungarian bureaucracy required all its employees to be members of a religion. Any religion would do. But Einstein, a non-believer, had stated on his application that he had no religion. When the difficulty was pointed out to him, he pragmatically filled in his religion as 'Mosaic' on the forms. He was also required to accept Austro-Hungarian citizenship, while retaining his Swiss citizenship, and was then officially appointed to the chair in January 1911, to take up his duties in March that year. The salary was the equivalent of 9,000 SF, enough for the family to rent a large apartment (with that new innovation, electric light) and employ a live-in maid.

On the move

Before taking up his post in Prague, Einstein (with Mileva) had visited Lorentz in Leiden. This was the first meeting between Einstein and the man who Abraham Pais has described as 'the one father figure in Einstein's life'. In an essay written in 1953 to mark the centenary of Lorentz's birth, Einstein wrote 'he meant more to me personally than anybody else I have met'.[6] Once in Prague, he continued to make new scientific friends, most notably Paul Ehrenfest, a Viennese physicist who shared Einstein's 'Jewish atheism' but unlike Einstein refused to compromise by pretending to a religious affiliation. At the time, he had been based in St Petersburg, where they were less scrupulous about such things, but visited Prague and met Einstein on a trip round the academic centres of Europe looking for a more congenial post. Ehrenfest soon succeeded Lorentz as professor in Leiden, much to Einstein's delight. Lorentz stayed on in a part-time post, giving Einstein a double reason for future visits to the Dutch city, which he did almost every year when he was based in Berlin, where he completed the General Theory. But the road from Prague to Berlin still had another turn to take.

Prague was far from being a quiet place to settle down to research in 1911. Although only about 5 per cent of the population of the city were German-speaking, they dominated cultural and political life, which inevitably caused friction with the bulk of the Czech population. In a further complication, about half of the Germans were Jewish and subject to other kinds of spoken and unspoken prejudice. In the last years of the Austro-Hungarian Empire Prague was a city where the German minority regarded themselves as socially superior to the Czech majority, the Czechs hated the Germans, and both communities disliked the Jews. Strange gypsy-like women from the Balkans were almost beneath contempt, but might just be tolerated if married to a respectable professor. In a letter to Besso, written soon after he arrived in Prague, Einstein said: 'My position and my institute give me much joy, only the people are so alien to me.'[7]

Like the rest of the Austro-Hungarian Empire, in the years after 1910 Prague was an accident waiting to happen. It didn't help that the city was filthy, and the water unfit to drink, which was not good for the baby. Einstein was to an extent insulated from all this by his absorption in his work, but it cannot have been comfortable for Mileva. She loathed Prague, resented being left out of the scientific discussions between Einstein and his colleagues, and was left to brood on her own when Einstein embarked on a series of scientific travels around Europe.

The most important of these was to attend a scientific conference, known as a Solvay Congress, in Brussels at the end of October 1911. This was the first of what became a long series of such events, sponsored by a wealthy Belgian chemist, Ernest Solvay. The theme of the first Solvay Conference (or Congress) was 'the quantum problem', and Einstein presented a paper on the application of quantum physics to the theory of specific heat. The details do not matter here, but his conclusion is as dramatic now as it was then: 'These discontinuities, which we find so distasteful in Planck's theory, seem really to exist in nature.'

In other words, quanta are real. As Lorentz pointed out at the time, this 'seems in fact to be totally incompatible with Maxwell's equations'. Einstein insisted that there had to be a way to reconcile the wave and particle theories of light: 'In addition to Maxwell's electrodynamics ... we must also admit a hypothesis such as that of quanta.'

As well as attending major conferences such as this, Einstein was often invited to give talks at universities around Europe. He also seems to have regarded the Prague post from the outset as merely a stepping stone in his career, and these visits provided a useful shop window for him to display his scientific wares. Mileva, increasingly suffering from depression, stayed in Prague with the children, and Einstein seems to have been glad to get away from her, as much as to spread the word about his ideas. It was against this background that on a visit to Berlin in Easter 1912 Einstein renewed acquaintance with a cousin, Elsa, who was three years older than Albert.

Elsa Einstein, as she was originally and would become again, was even more closely related to Albert than the term 'cousin' suggests. She was indeed the daughter of the sister of Albert's mother, but in addition her father Rudolf Einstein was the first cousin of Albert's father. She had been married, briefly, had two daughters and, although now divorced, was known by her married name of Löwenthal. Elsa and Albert had sometimes played together as children and once went to the opera together when a little older, but had not seen each other in years.

Elsa was in many ways the opposite of Mileva – a homely woman who enjoyed cooking and homemaking, and wanted nothing more than a comfortable life with a man who was a good provider. As for Albert, by now he was not looking for passion (though he never seems to have been averse to a dalliance) but craved a quiet domestic home life where he could be looked after and left to get on with his work.

Whatever the details, some kind of spark was struck between the two of them that spring. On his return to Prague, Einstein initially kept up a flow of correspondence with Elsa. In one letter he wrote: 'I have to have someone to love, otherwise life is miserable,' adding 'and this someone is you.' But then he had qualms. In May 1912 he wrote again to Elsa, this time saying that: 'It will not be good for the two of us, as well as for the others, if we form a closer attachment. So, I am writing to you today for the last time.'

As so often with Einstein's personal life, there was a professional reason associated with this change of heart. His attempts to find somewhere more congenial than Prague to work seemed to be bearing fruit, offering another 'last chance' to save his marriage. For some time, the University of Utrecht had been trying to recruit him, but Einstein had kept them on hold since he had an even better iron in the fire. In June 1911, the status of his alma mater, the ETH in Zurich, had been upgraded, and new professorships created. Even before he had left for Prague,

Einstein had promised that he would not accept an offer from another institution without giving the ETH a chance to find a niche for him. But the bureaucratic wheels in Zurich were turning painfully slowly, with doubts once again being expressed in some quarters about his ability as a teacher. Einstein kept the pot boiling in Utrecht in a deliberate attempt to stir the ETH into action, even asking the Dutch to delay announcing the fact that they had already found another candidate, Peter Debye. But eventually, aided by letters of recommendation from Marie Curie and Henri Poincaré, Einstein got the job. He would return to Zurich, and the ETH, in July 1912, at the end of the academic year in Prague. Einstein wrote to a friend with the news: 'Great joy about it among us old folks and the two bear cubs.' There was less joy at the University of Prague, but nothing they could do about it.

The post at the ETH could, perhaps *should*, have been a job for life. The family were able to afford a six-roomed apartment in a good part of town, they had friends and they fitted in. But Mileva, who should have been the chief beneficiary, developed rheumatism, which made it painfully difficult for her to go out during the winter, and became increasingly depressed in spite of the efforts of her friends to cheer her up. Einstein was never good at understanding the difficulties of others, let alone making a real effort to help them overcome such problems. And then Elsa came back into the reckoning.

In spite of breaking off his correspondence with her,

Einstein had thoughtfully given Elsa his new address in Zurich, and she sent him a present for his 34th birthday in March 1913, adding a request for a photograph of him and disingenuously asking advice on a good book on relativity theory. The correspondence was renewed, on much the same terms as before. At which point, in July 1913, two of the pillars of the German scientific establishment, Max Planck and Walther Nernst, arrived in Zurich with the proverbial offer that could not be refused.

The package put together by Planck and Nernst included a specially created professorship at the University of Berlin that need involve no teaching at all, if Einstein so wished it, together with membership of the prestigious Prussian Academy of Sciences (the youngest member of that august body), an enhanced salary and no administrative duties. It was a glorious offer, except for the impact such a move would surely have on Mileva. Einstein told Lorentz that: 'I could not resist the temptation to accept a position in which I am relieved of all responsibilities so that I can give myself over completely to rumination.' But that wasn't the whole story. In a letter to Heinrich Zangger, only released to the world in 2006, Einstein said that Elsa 'was the main reason for my going to Berlin'. If so, she has a lot to answer for, both good and bad.

Having just escaped from Prague back to a country and a city she felt comfortable in, it is easy to imagine Mileva's reaction when Albert told her about the job in Berlin. It was to prove the last straw, and although in the spring of 1914 Mileva did travel to Berlin with the children to join her husband, who had gone on ahead (not least to spend time with Elsa), it wasn't long before she returned to Switzerland, taking her sons with her. The final break came on 29 July 1914, when Einstein came to see off Mileva and the boys – who had already left his apartment and had been staying with friends in Berlin for several weeks – on the morning train to Zurich. Einstein cried like a child, particularly distressed at the loss of his children.

So Einstein was living alone in a large apartment in Berlin, but with Elsa and his own mother not far away, starting a new life when the First World War broke out. Even before the end of 1913, partly as a retreat from his domestic troubles, he had put the problems of quantum theory to one side and returned with full force to the attempt to generalise his theory of relativity. He now did so with an intensity which led to the breakthrough of 1915, while Elsa, hoping for the day when he would actually divorce Mileva and marry her, bided her time and kept his domestic affairs under control.

Einstein worked obsessively, slept only when he was exhausted, forgot to eat and neglected personal hygiene. The result was that by 1915 he completed his General Theory and presented it to the Prussian Academy. It would also, as we shall see, lead to a breakdown in his health in 1917, when his life was probably only saved by Elsa's attention.

First steps

For four years after his realization that 'if a person falls freely he will not feel his own weight', Einstein made no real effort to develop his ideas about non-uniform motion and gravity, even though he was convinced that this insight would enable him 'to extend or generalize the concept of relativity to apply to accelerated systems ... and in so doing, I expected that I would be able to resolve the problem of gravitation at the same time'.[8] The puzzles of quantum physics seemed more urgent and required his full attention. But in 1911, he told his friend Michael Besso that he was tired of quantum physics and was going to concentrate on developing his general theory of motion and gravity. Before we look at how he did so, though, it's worth spelling out just why a theory of accelerated objects is also a theory of gravity. The explanation is disarmingly simple, although nobody before Einstein spotted it.

Einstein used to describe things in terms of a falling lift, appropriate for the technology of his day. But these days it makes more sense to talk about the behaviour of objects inside a windowless spaceship. If the rockets are turned off and the spaceship is moving through space at a constant velocity, everything floats around inside the cabin, weightless. Under these conditions, if a beam of light is shone across the cabin from one wall to the other, it will travel in a straight line and make a spot on the opposite wall the same height above the floor as the height of the light source. The obvious light source to use would be a laser, and it is worth mentioning that the physics under-lying lasers was one of the things Einstein also worked out, just after completing the General Theory.

What happens if the rockets are turned on and the spaceship is accelerating?* Everything falls to the floor, and it feels to any occupants exactly the same as having weight. But what happens to the light beam? During the time the beam takes crossing the cabin, the speed of the spaceship has increased, moving the spaceship a tiny bit ahead of the light beam, so the spot made on the wall is a little lower than when there is no accelera-tion. It will look to any occupants of the spacecraft as if the light beam has been bent. Einstein realised (without ever having seen a spaceship) that in order for gravity and acceleration to be equivalent to one another – his insight from the patent office days – exactly the same thing must happen if the spaceship is sitting on the launch pad on Earth. With no windows, the occupants would not be able to tell if their weight was caused by acceleration or by gravity. But they *would* be able to tell if the light beam went in a straight line across the cabin when it was sitting on the ground. So, Einstein reasoned (from his equivalent thinking about falling lifts, rather than spaceships), gravity must bend light by just the right amount to make acceleration and gravity pre-cisely equivalent. This was his jumping off point when

---

* These are, of course, hypothetical silent rockets that make no vibration!

he returned to intense concentration on the search for a General Theory in 1911, during his time in Prague.

The first fruits of this effort came in a paper published in the *Annalen der Physik* in 1911, in which he calculated that 'a ray of light going past the sun would undergo a deflection of 0.83 seconds of arc'.* The exciting thing about this calculation was that it offered the prospect of testing the theory. Usually, we cannot see light passing close by the Sun, because of the glare of the Sun's own light. But during a solar eclipse, when the light from the Sun is blocked by the Moon, it is possible to see distant stars, far beyond the Sun, by light which skims past the edge of the Sun on its way to us. (It also skims past the Moon, but the Moon is far too small to produce a measurable effect on the light; the Sun is 27 million times more massive than the Moon.) If the light rays were deflected as Einstein predicted, these stars would appear in a slightly different place on photographic plates taken during an eclipse to their positions in photographs taken when the Sun was not in the line of sight. As Einstein put it: 'It would be a most desirable thing if astronomers would take up the question.'

The challenge was taken up by Erwin Freundlich, an astronomer at the Berlin University Observatory, who knew that there would be a solar eclipse visible from the Crimea on 21 August 1914. Einstein was so enthusiastic

---

* This number was later revised to 0.85 seconds of arc.

that he offered to help to fund an expedition to make the necessary observation, although in the end the money came from the Krupp Foundation. But the timing was disastrous. Freundlich and two colleagues left for the Crimea on 19 July 1914. On 1 August, as part of the complicated political web that led to the First World War, Germany declared war on Russia. Freundlich and his colleagues were taken prisoner by the Russians, and their equipment confiscated. Hardly surprisingly, as Germans carrying cameras and surveying equipment they were suspected of spying and were lucky to be returned to Germany a few weeks later as part of a prisoner exchange. But there was a silver lining, at least as far as the General Theory was concerned. Einstein's calculation from 1911 was wrong. If the expedition had been a success, it would have found a different deflection of light, providing a blow to Einstein's prestige (not that he would have cared) and perhaps discrediting the theory. But by the summer of 1914, amid the turmoil of his move to Berlin, the final breakup of his marriage and the outbreak of war, Einstein was already thinking along new lines.

'Everybody knows' that light travels in straight lines, so how could a light ray be bent? One possible answer lay in the application of a different kind of geometry to the problem, a geometry in which straight lines are not straight in the everyday sense, where parallel lines can cross or diverge, and where the angles of a triangle do not add up to 180 degrees. The familiar geometry in which parallel lines never meet or diverge, the angles of a triangle add up to 180 degrees, and so on is known as Euclidean geometry, after the Greek mathematician who spelled out the ground rules; it describes the geometry of a flat surface, in the everyday sense of the word 'flat'. This is the kind of geometry Minkowski used in his geometrisation of the Special Theory. Alternative geometries are therefore known as non-Euclidian geometries, and in the 19th century there had been a great deal of work on non-Euclidean geometries, largely as an abstract exercise in mathematics (although, as it happens, the geometry of the curved surface of the Earth is non-Euclidean). But Einstein did not know much about them, in spite of his earlier dilettante dabbling with Poincaré's work, and he knew even less about how to work with the corresponding equations. Indeed, it is a fact which still has the power to surprise that although Einstein had great insight into the physical nature of the world, he was always, by the standards of top-flight physics, rather weak at maths. But he knew a man who was a whizz at maths – his old friend Marcel Grossmann. One of Einstein's first actions on returning to Zurich in 1912 was to call on Grossmann for help in developing a set of mathematical equations – what physicists call a 'field theory' – that would enable him to write down the laws which describe the workings of gravity, or the gravitational field, in the way that Maxwell's equations describe the workings of the electromagnetic field. Grossmann, who did indeed know the history of non-Euclidean geometry, was only too pleased to help.

What Einstein should have known

The first person to go beyond Euclid and to appreciate the significance of what he was doing was the German Carl Friedrich Gauss, who was born in 1777 and had completed all of his great mathematical discoveries by 1799. But because he didn't bother to publish many of his ideas, non-Euclidean geometry was independently discovered by the Russian Nikolai Ivanovich Lobachevsky, who was the first to publish a description of such geometry in 1829, and by a Hungarian, János Bolyai. They all hit on essentially the same kind of 'new' geometry, which applies on what is known as a 'hyperbolic' surface, which is shaped like a saddle, or a mountain pass. On such a curved surface, the angles of a triangle always add up to less than 180 degrees, and it is possible to draw a straight line and mark a point not on that line, through which you can draw many more lines, none of which crosses the first line and all of which are, therefore, parallel to it.

But it was Bernhard Riemann, a pupil of Gauss, who comprehensively put across the notion of non-Euclidean geometry in the 1850s, and who realised the possibility of yet another variation on the theme – the geometry that applies on the closed surface of a sphere (including the surface of the Earth). In spherical geometry, the angles of a triangle always add up to more than 180 degrees, and although all 'lines of longitude' cross the equator at right angles and must therefore all be parallel to one another, they all cross each other at the poles.

Riemann, who had been born in 1826, entered Göttingen University at the age of twenty and learned his mathematics initially from Gauss, who had turned 70 by the time Riemann moved on to Berlin in 1847, where he studied for two years before returning to Göttingen. He was awarded his doctorate in 1851, and worked for a time as an assistant to the physicist Wilhelm Weber, an electrical pioneer whose studies helped to establish the link between light and electrical phenomena, partially setting the scene for Maxwell's theory of electromagnetism.

The accepted way for a young academic like Riemann to make his way in a German university in those days was to seek an appointment as a *Privatdozent*. In order to demonstrate his suitability for such an appointment, the applicant had to present a lecture to the faculty of the university, and the rules required the applicant to offer three possible topics for the lecture, from which the professors would choose the one they would like to hear. It was also a tradition, though, that although three topics had to be offered, the professors always chose one of the first two on the list. The story is that when Riemann presented his list for approval, it was headed by two topics which he had already thoroughly prepared, while the third, almost an afterthought, concerned the concepts that underpin geometry.

Riemann was certainly interested in geometry, but apparently he had not prepared anything along these lines at all, never expecting the topic to be chosen. But Gauss, still a dominating force in the University of Göttingen even in his seventies, found the third item on Riemann's list irresistible, whatever convention might dictate, and the 27-year-old would-be *Privatdozent* learned to his surprise that he would have to lecture on it to win his spurs.

Perhaps partly under the strain of having to give a talk he had not prepared and on which his career depended, Riemann fell ill, missed the date set for the talk and did not recover until after Easter in 1854. He then prepared the lecture over a period of seven weeks, only for Gauss to call a postponement on the grounds of ill health. At last, the talk was delivered on 10 June 1854. The title, which had so intrigued Gauss, was 'On the hypotheses which lie at the foundations of geometry'.

In that lecture – which was not published until 1867, the year after Riemann died – he covered an enormous variety of topics, including a workable definition of what is meant by the curvature of space and how it could be measured, the first description of spherical geometry (and even the speculation that the space in which we live might be gently curved, so that the entire Universe is closed up, like the surface of a sphere, but in three dimensions, not two), and, most important of all, the extension of geometry into many dimensions with the aid of algebra. Crucially, Riemann described mathematically the nature of a surface over which the geometry varies from place to place: flat in some places, hyperbolic in other places, spherical elsewhere.

Although Riemann's extension of geometry into many dimensions was the most important feature of his lecture, the most astonishing, with hindsight, was his suggestion that space might be curved into a closed ball. More than half a century before Einstein came up with the General Theory of Relativity – indeed, a quarter of a century before Einstein was even born – Riemann was describing the possibility that the entire Universe might be contained within what we would now call a black hole. 'Everybody knows' that Einstein was the first person to describe the curvature of space in this way – and 'everybody' is wrong.*

Of course, Riemann got the job. Gauss died in 1855, just short of his 78th birthday, and less than a year after Riemann gave his classic exposition of the hypotheses on which geometry is based. In 1859, on the death of Gauss's successor, Riemann himself took over as professor, just four years after the nerve-wracking experience of giving the lecture upon which his job as a humble *Privatdozent* had depended.

Riemann died of tuberculosis at the age of 39. If he had lived as long as Gauss, however, he would have seen his intriguing mathematical ideas about multi-dimensional

---

* Doubly wrong, as we shall see.

space begin to find practical applications in Einstein's new description of the way things move. But Einstein was not even the second person to think about the possibility of space in our Universe being curved, and he had to be set out along the path that was to lead to the General Theory of Relativity by mathematicians more familiar with the new geometry than he was. Chronologically, the gap between Riemann's work and the birth of Einstein is nicely filled by the life and work of the English mathematician William Clifford, who lived from 1845 to 1879 and who, like Riemann, died of tuberculosis. Clifford translated Riemann's work into English and played a major part in introducing the idea of curved space and the details of non-Euclidean geometry to the English-speaking world. He knew about the possibility that the three-dimensional Universe we live in might be closed and finite, in the same way that the two-dimensional surface of a sphere is closed and finite, but in a geometry involving at least four dimensions. This would mean, for example, that just as a traveller on Earth who sets off in any direction and keeps going in a straight line will eventually get back to their starting point, so a traveller in a closed universe could set off in any direction through space, keep moving straight ahead and eventually end up back at their starting point.

But Clifford realised that there might be more to space curvature than this gradual bending encompassing the whole Universe. In 1870, he presented a paper to the Cambridge Philosophical Society in which he described the possibility of 'variation in the curvature of space' from place to place, and suggested that: 'Small portions of space are in fact of nature analogous to little hills on the surface [of the Earth] which is on the average flat; namely, that the ordinary laws of geometry are not valid in them.' In other words, still seven years before Einstein was born, Clifford was contemplating local distortions in the structure of space – although he had not got around to suggesting how such distortions might arise, nor what the observable consequences of their existence might be, and the General Theory of Relativity, as we shall see, actually portrays the Sun and stars as making dents, rather than hills, in spacetime, not just in space.

Clifford was just one of many researchers who studied non-Euclidean geometry in the second half of the 19th century – albeit one of the best, with some of the clearest insights into what this might mean for the real Universe. His insights were particularly profound, and it is tempting to speculate how far he might have gone in pre-empting Einstein, if he had not died eleven days before Einstein was born. So Einstein was following an established tradition when, with Grossmann's help, he picked up the threads and completed his masterwork.

The masterwork

In describing the Special Theory in geometric terms, a crucial property is the velocity of an object. Velocity involves both speed and direction, and in physics such directional entities are called 'vectors'. In the curved space-times described by Riemannian geometry, the equivalents of vectors are called 'tensors', and they are more complicated in that they involve more numbers (there are more components to a tensor than there are to a vector). This is what makes calculations involving tensors tedious, although, once you know the rules, they are not difficult in the sense of requiring great brain power. Working out the rules is the clever bit.

The distances between points in Riemannian space are calculated in terms of a particular kind of tensor, known as the 'metric tensor'. The metric tensor that applies to curved four-dimensional spacetime has been described as a 'vector on steroids' and has sixteen components, although only ten of them are completely independent of one another. It is usually written as $g_{\mu\nu}$ and pronounced 'gee mu nu'. This expression means as much (or more!) in the General Theory as $E = mc^2$ does in the Special Theory.

Its essence can be gleaned quite easily using Minkowski's ideas about spacetime. When I mentioned earlier using coordinates to describe a triangle on a sheet of graph paper, I took it for granted you would be thinking of a flat sheet of paper. Einstein's Special Theory describes the way things move about in what is called 'flat space-time'. Einstein's General Theory of Relativity describes how things move in *curved* spacetime, and the curvature in spacetime is caused by the presence of matter. For lines drawn on the surface of a sphere, for example, the angles of a triangle don't add up to 180 degrees, and lines that start out parallel to one another (like the lines of longitude crossing the equator) can end up crossing each other (in this case, at the North and South poles).

The usual example is to think of a stretched rubber sheet, like a trampoline. This is flat, and if you roll marbles across it they travel in straight lines. Now imagine dumping a heavy weight, like a bowling ball, on the stretched sheet. It makes a dent – it curves 'spacetime' – and if you roll marbles across the curved surface they will follow curved paths past the bowling ball. Einstein's key insight was that gravity depends on the curvature of spacetime, telling material objects how to move, while the presence of material objects is what makes spacetime curved. He later explained his insight to his son Eduard: 'When a blind beetle crawls across the surface of a curved branch, it doesn't notice that the track it has covered is indeed curved. I was lucky enough to notice what the beetle didn't notice'.[9] But it took him three more years to turn that insight into a rigorous mathematical theory.

Einstein's approach rested on two legs. One was physical. He knew that his theory of gravity must reproduce all the results of Newton's theory in the case of weak gravitational fields, that it must contain the basic laws of 'classical' (that is, Newtonian) physics such as conservation of energy and momentum, and, his own Big Idea, that experiments carried out in an accelerating laboratory (or frame of reference) must give the same results as equivalent experiments carried out in a stationary laboratory in a gravitational field. The second leg was mathematical, using tensor theory to find an equation describing gravity (a field equation, or set of field equations) that is independent of arbitrary changes in the space and time coordinate systems (equivalent to changing the meridian we measure longitude from, from Greenwich to, say New York; wherever you measure from the distances between places stay the same). Such a theory is said to be covariant.

Einstein's notebooks show that by the end of 1912, when he was still in Zurich, he had found a field equation that satisfied the mathematical requirements, but did not seem to produce the same results as Newton for weak fields. Never a great mathematician, but a firm believer in physical intuition, this led Einstein to concentrate on the physical side of the problem in the months that followed, pushing the mathematics aside. This turned out to be a mistake; the equation he rejected in 1912 was very nearly the solution he had been looking for.

Starting out from the physical requirements, in the spring of 1913 Einstein and Grossmann produced what they called an 'Outline of a Generalised Theory of Relativity and a Theory of Gravitation', known now as the '*Entwurf*', the German word for 'outline'. The snag was, this theory was not covariant – people in different accelerated frames of reference would not see the universe in the same way. But it did make some interesting predictions.

Among other things, it predicted that the orbit of Mercury would change over time in a regular way. There was an observed 'advance in the perihelion of Mercury' that could not be explained by Newtonian theory. The *Entwurf* did imply such a phenomenon, but Michele Besso showed that the predicted effect, although arguably better than nothing, was much too small to explain the observations. This echoes, with hindsight, the problem with Einstein's light bending prediction of 1911: in the right direction, but too small. As I have mentioned, if Erwin Freundlich's eclipse expedition of 1914 had been successful, Einstein would have been proved wrong!

Einstein struggled with the *Entwurf* idea during the months that saw turmoil in his domestic life and the move to Berlin in the spring of 1914. Living alone (but near to Elsa) in a seven-roomed apartment, he was free to work long and irregular hours, surrounded by piles of paper but the bare minimum of furniture, refining the theory and attempting to answer criticisms of it. By the spring of 1915, he was no longer referring in his notes to 'a generalised theory' but to 'the general theory', and he explained his progress so far in a series of lectures at the University of Göttingen in the summer of 1915. The lectures were a great success, and Einstein got on like a house on fire with the professor of mathematics there, 53-year-old David Hilbert, who enthusiastically espoused Einstein's ideas. Or at least, one leg of them. As a mathematician, Hilbert was intrigued by the second leg of the theory, which Einstein had been neglecting for the past couple of years. He set out to find the required covariant field equation, without worrying too much about the physical requirements. And while he did so (making more progress than Einstein was aware of at the time), Einstein was finally forced to abandon the *Entwurf* approach.

In October 1915, Einstein finally convinced himself that the problems with covariance, with the orbit of Mercury and with the description of rotating systems made the *Entwurf* unworkable. Physical insight had, for once, let him down. So he went back to his Zurich notebooks and picked up the tensor leg of the theory from where he had left off. In a manner reminiscent of the way that problems can sometimes be solved by sleeping on them, when he looked again at the calculations from 1912 he saw almost immediately where he had gone wrong, and how with a relatively minor tweak he could come up with properly covariant field equations. But that still implied a great deal of calculation to get all his beans in a row. This led to a furious burst of work which he was able to present in instalments (lectures based on scientific papers), while still developing the theory, to four successive weekly meetings of the Prussian Academy of Sciences, starting on Thursday, 4 November 1915. Even when he gave the first lecture, though, he had not yet worked through to the fully covariant field equations; this was very much a work in progress.

Einstein knew that Hilbert was working on the

problem, but not how much progress he had made. So he sent him a copy of the 4 November lecture, asking what Hilbert thought of this new approach. The lecture of 11 November, which still didn't effectively solve the covariance problem, went off to Göttingen in the same way. Einstein was alarmed to receive a letter from Hilbert informing Einstein that he was on the brink of a solution to 'your great problem' and that if Einstein could come to Göttingen the following Tuesday, 16 November, he would be glad to 'lay out my theory in very complete detail'. He even reminded Einstein of the times of the trains from Berlin to Göttingen. But what must have been the most alarming news from Hilbert was left for a postscript – 'as far as I understand your new paper, the solution given by you is entirely different from mine'.

Einstein had no intention of going to Göttingen with his own theory still incomplete and Hilbert possibly about to pre-empt him. On 15 November he wrote back declining the invitation, complaining of stomach pains and fatigue (he was genuinely ill as a result of overwork and a tendency not to bother with things like eating when he was busy), and asking for a copy of Hilbert's theory 'to mitigate my impatience'.

Flinging himself back into his calculations, Einstein made a great discovery, just in time for his third lecture on 18 November. Reworking the calculation of the orbit of Mercury with the revised version of his theory gave the right answer! The theory now predicted a shift in the perihelion of 43 seconds of arc per century, exactly matching observations. Einstein was so excited that he suffered heart palpitations and had to take a rest. But this wasn't all. The same revision to the theory gave a new prediction for the bending of light by the Sun, not 0.85 seconds of arc as he had previously calculated, but exactly twice as much, 1.7 seconds of arc. This was less exciting, because in the absence of a convenient solar eclipse there were no observations to match the theory up with, and wouldn't be for several years. But both results were included in the 18 November lecture.

Einstein's excitement that day was slightly dampened by the arrival of a letter from Hilbert containing a copy of his own work, which, in spite of the disclaimer in the previous letter, turned out to be along very similar lines to Einstein's work. He replied (before giving his lecture!) with a carefully worded letter, clearly intended to establish his priority and worth quoting at length:

> The system you furnish agrees – as far as I can see – exactly with what I found in the last few weeks and have presented to the Academy … Three years ago with my friend Grossmann I had already taken into consideration the only covariant equations, which have now been shown to be the correct ones. We had distanced ourselves from it, reluctantly, because it seemed to me that the physical discussion yielded an incongruity with Newton's law. Today I

am presenting to the Academy a paper in which I derive quantitatively out of general relativity, without any guiding hypothesis, the perihelion motion of Mercury. No gravitational theory has achieved this until now.

Hilbert replied by return, offering his congratulations. But he also, on 20 November, sent his version to a science journal based in Göttingen for publication. Just how much – or how little – Einstein was influenced by Hilbert's work we will never know for sure. But on 25 November 1915, in his fourth lecture to the Prussian Academy, under the title 'The Field Equations of Gravitation', Einstein presented the covariant field equations that are the basis of the General Theory – the $g_{\mu v}$ equations that mean as much to the General Theory as $E = mc^2$ does to the Special Theory. The General Theory of Relativity was essentially complete.*

Was it all Einstein's own work? The evidence suggests it was, although Hilbert was within a hair's breadth of getting there first. Although Hilbert's published paper was dated 20 November, five days before Einstein's fourth lecture, a set of page proofs survive which shows that he made some corrections to the paper before finally

---

* The presentations to the Academy were published bit by bit as the talks went along, but the definitive formal paper putting it all together, 'The Foundation of the General Theory of Relativity', appeared in 1916 in the *Annalen der Physik*, volume 49, p. 79.

returning it to the publisher on 16 December. Those corrections were necessary because the original equations were not fully covariant – and they seem to have been made in the light of what Einstein said on 25 November. In other words, Hilbert corrected his work to match Einstein, not the other way around. As Hilbert himself put it when presenting the equations, they were 'first introduced by Einstein'. And it was, of course, Einstein's theory that Hilbert was working on, as Hilbert always acknowledged. Einstein mailed printed copies of his four lectures to his colleagues (full publication took place in 1916), describing them as containing 'the most valuable discovery of my life'. He was right. At the age of 36 he had changed our view of the Universe more profoundly than anyone except Newton. To explain why, I shall describe the legacy of the General Theory, before returning to Einstein's life and later work.

4 Legacy

Black holes and timewarps; Beyond reasonable doubt; Making waves; The Universe at large

The first person to sit up and take notice of Einstein's General Theory – apart from the people he had been discussing it with – was Karl Schwarzschild, the former director of the Potsdam Observatory, who was now serving with the German army on the Eastern Front, calculating the trajectory of artillery shells. By early 1916 he had not only received copies of Einstein's presentations to the Prussian Academy, but had found an exact solution to the equations, which he mailed to Einstein, who in turn presented it to the next meeting of the Academy on 16 January.

Schwarzschild's breakthrough was more of an achievement than it might seem to a non-mathematician. Einstein had found a description of gravity in terms of a set of ten coupled nonlinear partial differential equations (the full version of the $g_{\mu\nu}$), which on the face of things were very difficult to solve fully. It was one thing to apply them to a specific problem, such as the calculation of the orbit of Mercury, but quite another to get an exact solution which described the complete behaviour of spacetime near a mass. But that is what Schwarzschild had done. First, he solved the equations to provide a description of the behaviour of space everywhere outside a spherical, non-rotating mass, such as a star; then, in a second communication to Einstein, he provided a description of how things are inside such an object. Both results implied the possibility of what are now known as black holes.

Black holes and timewarps

Schwarzschild's solution to Einstein's equations (known today simply as the Schwarzschild solution) implied that if any mass were squeezed into a small enough volume, two things would happen. The space around the object would be bent round upon itself so that it would be sealed off from the outside world and nothing could escape (not even light; hence the later introduction of the term 'black hole'); and the matter inside this 'Schwarzschild radius' would collapse all the way down to a mathematical point, a singularity. The appearance of a singularity in a theory is usually a sign that something is wrong – that the equations that are being applied do not work under such extreme conditions. Einstein himself never believed that such Schwarzschild singularities could exist in the real world, or even that real objects could shrink within the appropriate Schwarzschild radius. But what the equations implied was that the Sun would become a black hole if it were squeezed within a radius of 2.9 kilometres, and that the Earth would do so if squeezed to the size of a large bean, with a radius of 0.88 centimetres. The bigger the mass involved, the bigger the Schwarzschild radius and the easier it would be to make a black hole.

It took decades for these ideas to even begin to be taken seriously as applying to real objects in the Universe, and Schwarzschild, who contracted a rare skin disease and died on 11 May 1916, never knew what he had started. The first hint that the Schwarzschild radius might at least be approached by collapsing objects in the real world came when Subrahmanyan Chandrasekhar, an Indian physicist based in England, applied the then-new understanding of quantum mechanics to calculate the fate of a star that had run out of nuclear fuel and could no longer generate heat in its interior to provide the pressure to hold itself up against the pull of gravity. It was thought that a star at the end point of its life would become a solid object with the atoms jammed tightly together, something about the size of the Earth but containing about as much mass as the Sun. But at the beginning of 1930 Chandrasekhar showed that if such a white dwarf star had more than about 1.5 times the mass of the Sun then the atoms themselves would be crushed and the star would collapse even further. His ideas were not widely accepted at the time, but in Russia Lev Landau suggested that under such extreme conditions electrons and protons would be fused to make neutrons, and the star (or at least, its core) would become a ball of neutrons like a huge atomic nucleus. The idea was born of a neutron star, containing about as much mass as the Sun contained in a volume about the same as that of Mount Everest. But it turned out that even this was not the last word. In 1938, Robert Oppenheimer and George Volkoff, in America, found that even a neutron star could not hold itself up against the pull of gravity if its mass exceeded a certain amount, now known as the 'Oppenheimer-Volkoff limit'. This is about three times the mass of the Sun. For greater masses, they wrote in a paper published in 1939, 'the star will continue to contract indefinitely, never achieving equilibrium'.

Oppenheimer discussed the implications in another paper, published later in 1939:

> When all the thermonuclear sources of energy are exhausted a sufficiently heavy star will collapse. Unless fission due to rotation, the radiation of mass, or the blowing off of mass by radiation, reduce the star's mass to the order of that of the sun, this contraction will continue indefinitely ... the radius of the star approaches asymptotically its gravitational radius ... The total time of collapse for an observer co-moving with the stellar matter is ... of the order of a day.[1]

Or in everyday language, the star collapses into a black hole in a matter of a few hours.

But hardly anyone believed that neutron stars, let alone black holes, really existed. Surely there must be some law of nature to prevent such an absurdity? There things stood until 1967, when the discovery of pulsars forced a rethink. The only workable explanation for these rapidly flickering radio sources turned out to be that they were neutron stars, spinning quickly and flicking beams of radio noise around like the beams of a lighthouse. If neutron stars were real, astronomers realised, so might black holes be. It is no coincidence that the revival of interest in black holes, and John Wheeler's coining of the name in December 1967, followed hot on the heels of the discovery of pulsars. Since then, it has become abundantly clear, especially from observations using X-ray telescopes orbiting the Earth, that black holes are not only real but far from rare. 'Stellar mass' black holes have been discovered orbiting ordinary stars and revealing their presence by the energy radiated by matter falling into the holes; and 'supermassive' black holes, millions of times as massive as our Sun and as big as the Solar System, have been identified by their similar (but bigger) energetic outbursts at the hearts of galaxies, including our Milky Way. Perhaps Einstein would have been astonished by all this. Or perhaps not, in view of some of his own ideas about what happens to space and time when they are distorted by matter.

What happens inside a black hole? The short answer is, nobody knows. But the equations give us some clues. A key feature of those equations is that it is possible to define a kind of interval in spacetime that is the same for all observers, however they are moving. In physical terms, relative motion may make time stretch and space shrink, so that a kind of average of the two stays the same. This is called a 'metric'. Different metrics apply in different kinds of curved spacetime, and the description of a black hole found by Schwarzschild can be referred to as the Schwarzschild metric. As early as 1916, the Austrian Ludwig Flamm noticed that the Schwarzschild solution to Einstein's equations actually describes not one black hole (using modern parlance) but two, connected by some kind of tunnel, which became known as a 'wormhole'. This led to some intriguing speculation by physicists. If an electron, or some other particle, existed literally at a point in space, then the spacetime around it would be described by the Schwarzschild metric. Could it be that particles such as electrons actually existed in pairs, perhaps widely separated in space, but connected by wormholes? The appeal of this idea was that it raised the possibility that particles – matter – might be described entirely in terms of curved spacetime. This idea turned out to be a blind alley.* But it did lead Einstein, working with Nathan Rosen at Princeton in the mid-1930s, to investigate the mathematical description of larger versions of such wormholes, which became known to relativists as 'Einstein-Rosen bridges'.

An Einstein-Rosen bridge is like a tunnel through

---

* But there is a modern counterpart, known as string theory.

spacetime, linking two flat regions of space which may be far apart – even on opposite sides of the Universe. If such an object existed – and it is a big *if* – and if the entrances were big enough, you could jump in one end of the wormhole and come out of the other. This is the basis of many a science fiction story, so Star Trek is part of Einstein's legacy. The big *if*, though, is related to the fact that even if such an object formed, it would collapse much more quickly than the time it would take to traverse it. But the idea was revived, 50 years later, when theorists realised what should have been obvious from the start – a tunnel through spacetime can take you to a different place *or* to a different time (or both!). The equations were telling us that if an Einstein-Rosen bridge could be kept open, it would operate as a time machine. There are immense practical difficulties (see my and Mary's book *Time Travel for Beginners*). But the bottom line is that according to Einstein's equations – the best description of the Universe at large that we have – time travel is not impossible, just very, very difficult. Could Einstein be wrong? Well, so far the general theory has passed every test, the latest as recently as 2013, with flying colours.

Beyond reasonable doubt

The most famous test of the General Theory came in 1919, when the light bending effect was observed during an eclipse of the Sun. But that story belongs in the next chapter, as a key event in Einstein's life. The explanation of the shift in the perihelion of Mercury was another contemporary proof that Einstein was right. But the most impressive proofs of the accuracy of his theory are part of his legacy, having been obtained after his death.

Two related kinds of test of the General Theory depend on a phenomenon known as the 'gravitational redshift'. The simplest way to understand how this arises is to go back to the image of a freely falling lift derived from Einstein's vision that 'if a person falls freely he will not feel his own weight'. If gravity and acceleration are equivalent, everything going on inside the lift must look the same to observers inside and outside the lift. Inside the lift, a pulse of light is emitted from a laser at the bottom, and proceeds vertically towards the roof. While it is on its way, an observer inside the lift measures its wavelength. The lift and everything it contains is in free fall, so the light wave travels at the speed of light and with unchanging wavelength. But if the pulse of light is observed through a window by someone outside the lift, sitting still on the surface of the Earth, because of the motion of the lift the light will be squashed to shorter wavelengths – a 'blueshift' (such an observer would have to have supersensitive instruments, or use a high-speed movie camera and play back the scene in slow motion). The only physical difference is that the outside observer is sitting still in a gravitational field. So the only way the two observers can measure exactly the same wavelength for the light pulse is if gravity causes a redshift, stretching the light, which exactly cancels out the squashing, producing the blueshift. This, of course, can all be described mathematically by Einstein's equations. The prediction is that for light near the surface of the Earth, the measured frequency of a laser beam should be shifted by one trillionth of 1 per cent for a difference in height of 100 metres. This is the gravitational redshift. Amazingly, in the 1960s physicists were able to carry out appropriate experiments to this precision.

The experiments were carried out by Robert Pound and colleagues at Harvard University. The Jefferson Tower of the physics building at Harvard has a lift shaft 22.5 metres tall, and Pound's team measured the change in wavelength (actually a blueshift, because the light is going down, not up) of radiation emitted at the top of the tower and recorded at the bottom of the tower to a precision of two parts in a thousand trillion. This required an emitter that produced a very precise wavelength (or frequency) of light, and a detector that could measure the wavelength with exquisite precision. The source was a sample of radioactive material, iron-57, which emits gamma rays with a very precise frequency, corresponding to a wavelength of 0.86 Ångstrom, or 0.086 nanometres. The same material will absorb gamma rays, but only if they have exactly the right frequency. So a detector incorporating iron-57 was placed at the foot of the shaft on a hydraulic lift which moved slowly downwards to produce a Doppler redshift, cancelling out the gravitational blueshift, and the speed of the lift at the point where so-called 'resonance' occurred and the gamma rays were absorbed was measured. The numbers are mind-boggling, but some idea of the precision of the experiment is shown by the fact that resonance occurred only when the hydraulic lift was moving downwards at two millimetres per hour. This was, of course, exactly in line with the predictions of the General Theory.

This gravitational influence affects not only measurements of length such as the wavelength of light, but also time, so that clocks in a stronger gravitational field should run more slowly than identical clocks in a weaker gravitational field. This is actually the same thing as the gravitational redshift, looked at from a different perspective. In effect, we measure the wavelength of light by counting how many wave peaks go past a fixed point in one tick of the clock. If the wave is redshifted, fewer peaks go past in the same time. But we would get exactly the same effect if the ticks of the clock were closer together, so there was less time for the waves to get past. The reverse is true for a blueshift. In our falling lift experiment, the observer on the ground says the light is redshifted, but the observer in the lift says the clock on the ground is running slow.

The experiments that tested this variation on the theme were more complicated than the Harvard lift experiment, since they involved moving clocks (on aeroplanes: the only way to get them to a high enough altitude to measure the effect), which meant that the effects described by both the Special Theory and the General Theory had to be taken into account. So it wasn't until 1971 that direct proof of gravitational time dilation was obtained.

The experimenters who took on this daunting task were Joseph Hafele, of Washington University, St Louis, and Richard Keating, of the US Naval Observatory in Washington, DC. If they had been able to charter a private jet to fly their ultra-precise atomic clocks in, life would have been reasonably simple. But government financial rules restricted their budget, so they had no choice but to fly their clocks around the world on commercial aircraft on scheduled flights. They couldn't even travel first class, but in economy, with the clocks strapped securely to the front wall of the passenger cabin and connected to the aircraft's power supply. The experiment involved flying a pair of clocks eastward, right around the world, between 4 and 7 October 1971, and westward around the world between 13 and 17 October. The two flights in opposite directions were necessary in order to make allowance for the rotation of the Earth. And, being commercial flights, the circumnavigations involved many stopovers, changes in speed, altitude and direction, all of which had to be carefully logged and fed into the calculations when the times recorded on the traveling clocks were compared with the times recorded on counterparts which had sat quietly at the US Naval Observatory while all this was going on.

The results were not as precise as the Harvard lift shaft experiment, but matched the predictions of the General Theory within the limits of experimental error. For the westward flight, the clock gained 273 nanoseconds (billionths of a second), compared with a prediction of 275 nanoseconds, two-thirds of this attributable to the gravitational blueshift resulting from being at altitude. The situation was more complicated for the eastward flight, where the time dilation effect of the Special Theory was expected to cause the clock to lose more than it would gain from the gravitational blueshift. The observed loss was 59 nanoseconds, but with a range of errors caused by inaccuracies in the flight data allowing for the possibility of anything from 39 to 79 nanoseconds; the prediction was 40 nanoseconds. Suggestive, but not on its own compelling.

But if anyone doubted the reality of the gravitational redshift, those doubts were laid to rest in June 1976, when a rocket-borne experiment, devised by Robert Vessot and Martin Levine and known as Gravity Probe A, soared into space to an altitude of 10,000 kilometres on a sub-orbital flight.* During the flight, the 'ticking' of an atomic clock in the payload was compared over a radio link with an identical clock on the ground, monitoring the way the changing Doppler shift resulting from the motion of the rocket relative to the ground interacted with the

---

* It actually took two years to analyse all the data, so the final results came out in 1978.

gravitational blueshift at altitude to produce an overall difference between the two clocks. The results matched the predictions of relativity theory to a precision of 70 parts per million, 7 thousandths of 1 per cent.

Another test of Einstein's theory, known as Gravity Probe B, took place at lower altitudes, but in orbit rather than on a sub-orbital flight. This was a long time in the preparation, and although it produced results towards the end of the first decade of the 21st century, it is close to my heart since it was the subject of one of the first articles I ever wrote for the magazine *New Scientist*, 40 years earlier, back in 1968 when I was a PhD student.

Gravity Probe B was also known as 'the Stanford weightless gyro experiment'. This neatly sums up what it was all about. The lead scientists on the project came from Stanford University, under the leadership of Francis Everitt, and it involved a set of gyroscopes floating weight-lessly in orbit around the Earth in an unmanned satellite. The aim of the experiment was to measure the effect on spacetime of the rotation of the Earth, which has been likened to the distortion in a bowl of maple syrup produced by twisting a spoon round in the middle of the syrup. It is known as the 'Lense-Thirring effect', because it was suggested by Austrian physicists Josef Lense and Hans Thirring two years after Einstein published his General Theory of Relativity. The effect is also known as 'frame dragging', and it is tiny, which is why it took so long to develop instruments sensitive enough to measure it and get them into space. Happily, all we have to worry about here are the results.

There were four gyros on board the satellite, which orbited the Earth at an altitude of 650 kilometres passing over both poles. It was launched on 20 April 2004, and the mission lasted for sixteen months. The launch had to be timed to a precision of one second to get the payload into the precise orbit needed for the experiment. But this was trivial compared with the precision of the experiment itself. The gyros, about the size of table-tennis balls, were the most perfectly round objects ever made. They were spherical to within 40 atoms and if one of them could have been expanded to the size of the earth, the biggest irregularity on the surface would be only 2.4 metres high. The spheres themselves were made of fused quartz, coated with an extremely thin layer of the metal niobium. They could never touch the walls of their container, so they were suspended with electric fields, and made to spin with a squirt of helium gas; the directions in which they were pointing (their spin axes) were monitored using magnetic fields produced by the niobium layer. Full results from the experiment appeared in the journal *Physical Review Letters* in 2011. They revealed that for all four gyroscopes the 'drift' in the direction the gyros were pointing matched the predictions of the General Theory. The measured frame-dragging drift rate was 37.2 milliarcseconds per year, with possible errors of ±7.2 mas/yr; the prediction was 39.2 mas/yr.

But matter doesn't only drag spacetime round with it. It can also make ripples in spacetime, and in 2013 one consequence of such gravitational radiation provided the most precise confirmation yet of the accuracy of Einstein's theory.

Making waves

Gravity Probe B confirmed that massive objects, such as the Earth or a star, drag spacetime around with them as they rotate. If they move back and forth, they can also generate waves in the fabric of spacetime, known as 'gravitational waves', or 'gravitational radiation', like the ripples you can make in a bowl of water by wiggling your finger about in it. The resulting ripples in the fabric of space are very weak, unless a very large mass is involved in fairly rapid motion. But the waves were predicted by Einstein in a paper published in 1916, where he showed that they should move at the speed of light. Physicists have been trying for decades (as yet unsuccessfully) to detect gravitational radiation using very sensitive detectors here on Earth, and plan to put even more sensitive detectors into space. But meanwhile absolute proof of the accuracy of his prediction has come from observations of compact objects far away in space – the latest, and most precise, of these observations being reported in 2013.[2] The objects involved are compact binary stars, systems in which one star orbits closely around another – or rather, where both stars orbit around their common centre of mass, like a whirling dumbbell or the twirling mace of a drum majorette. The first of these systems extreme enough to test Einstein's prediction was a 'binary pulsar', studied in the mid-1970s.

A binary pulsar exists when two neutron stars, one of which is a pulsar, are in orbit around one another, forming a binary star system. The term is also used to refer to a pulsar in orbit about any other star, for example, a white dwarf. More than twenty binary pulsars are now known, but astronomers reserve the term 'the binary pulsar' for the first one to be discovered, which is also known by its catalog number, as PSR 1913+16.

The binary pulsar was discovered in 1974 by Russell Hulse and Joseph Taylor, of the University of Massachusetts, working with the Arecibo radio telescope in Puerto Rico. This pulsar was at the time the most accurate clock yet discovered. What they found that summer was so important that in 1993 the pair received the Nobel Prize for their work on the binary pulsar.

The first hint of the existence of the binary pulsar came on 2 July, when the instruments recorded a very weak signal. Had it been just 4 per cent weaker still, it would have been below the automatic cutoff level built into the computer program running the search, and would not have been recorded. The source was especially interesting because it had a very short period, only 0.059 seconds, making it the second fastest pulsar known at the time. But it wasn't until 25 August that Hulse was able to use the Arecibo telescope to take a more detailed look at the object.

Over several days following 25 August, Hulse made a series of observations of the pulsar and found that it varied in a peculiar way. Most pulsars are superbly accurate clocks, beating time with a precise period measured to six or seven decimal places; but this one seemed to have an erratic period which changed by as much as 30 microseconds (a huge 'error' for a pulsar) from one day to the next. Early in September 1974, Hulse realised that these variations themselves followed a periodic pattern, and could be explained by the Doppler Effect caused by the motion of the pulsar in a tight orbit around a companion star. Taylor flew down to Arecibo to join the investigation, and together he and Hulse found that the orbital period of the pulsar around its companion (its 'year') is seven hours and 45 minutes, with the pulsar moving at a maximum speed (revealed by the Doppler effect) of 300 kilometers per second – one tenth of the speed of light – and an average speed of about 200 kilometers per second, as it zipped around its companion. The size of the orbit traced out at this astonishing speed in just under eight hours is about 6 million kilometres, roughly the circumference of the Sun. In other words, the average separation between the pulsar and its companion is about the radius of the Sun, and the entire binary pulsar system would neatly fit inside the Sun.

All pulsars are neutron stars; the orbital parameters

showed that in this case the companion star must also be a neutron star. The system was immediately recognised as an almost perfect test bed for the General Theory – and, indeed, for the Special Theory, as well. As I have explained, one of the key tests of the General Theory is the advance of the perihelion of Mercury. The equivalent effect in the binary pulsar (the shift in the 'periastron') would be about 100 times stronger than for Mercury, and whereas Mercury only orbits the Sun four times a year, the binary pulsar orbits its companion 1,000 times a year, giving much more opportunity to study the effect. It was duly measured and found to conform exactly to the predictions of Einstein's theory – the first direct test of the General Theory made using an object outside the Solar System. By feeding back the measurements of the shift into the orbital data for the system, the total mass of the two stars in the system put together was eventually determined to unprecedented accuracy, as 2.8275 times the mass of our Sun.

But this was only the beginning of the use of the binary pulsar as a gravitational laboratory in which to test and use Einstein's theory. Extended observations over many months showed that, once allowances were made for the regular changes caused by its orbital motion, the pulsar actually kept time very precisely. Its period of 0.05903 seconds increased by only a quarter of a nanosecond (a quarter of a billionth of a second) in a year, equivalent to a clock that lost time at a rate of only 4 per cent in a million years.

The numbers became more precise as the observations mounted up. For 1 September, 1974, the data were: period, 0.059029995271 sec; rate of increase, 0.253 nanoseconds per year; orbital period, 27906.98163 seconds; rate of change of periastron, 4.2263 degrees of arc per year.

The accuracy of the observations soon made it possible to carry out more tests and applications of the theory of relativity. One involves the time dilation predicted by the Special Theory of Relativity. Because the speed of the pulsar around its companion is a sizeable fraction of the speed of light, the pulsar 'clock' is slowed down, according to our observations, by an amount which depends on its speed. Since the speed varies over the course of one orbit (from a maximum of 300 km/sec down to 'only' 75 km/sec), this will show up as a regular variation of the pulsar's period over each orbit. And because the pulsar is moving in an elliptical orbit around its companion, its distance from the second neutron star varies. This means that it moves from regions of relatively high gravitational field to regions of relatively low gravitational field, and that its timekeeping mechanism should be subject to a regularly varying gravitational redshift.

The combination of these two effects produces a maximum measured variation in the pulsar period of 58 nanoseconds over one orbit, and this information can be fed back in to the orbital calculations to determine the ratio of the masses of the two stars. Since the periastron shift tells us that the combined mass is 2.8275 solar masses, the addition of these data reveals that the pulsar itself has 1.42 times the mass of our Sun, while its companion has 1.40 solar masses. These were the first precise measurements of the masses of neutron stars.

But the greatest triumph of the investigation of the binary pulsar was still to come. Almost as soon as the discovery of the system had been announced, several relativists pointed out that in theory the binary pulsar should be losing energy as a result of gravitational radiation, generating ripples in the fabric of spacetime that would carry energy away and make the orbital period speed up as the binary pulsar and its companion spiraled closer together as a result.

Even in a system as extreme as the binary pulsar, the effect is very small. It would cause the orbital period (about 27,000 seconds) to increase by only a few tens of a millionth of a second (about 0.0000003 per cent) per year. The theory was straightforward, but the observations would require unprecedented accuracy. In December 1978, after four years of work, Taylor announced that the effect had been measured, and that it exactly matched the predictions of Einstein's theory. The precise prediction of that theory was that the orbital period should decrease by 75 millionths of a second per year. By 1983, nine years after the discovery of the binary pulsar, Taylor and his colleagues had measured the change to a precision of 2 millionths of a second per year, quoting the observed value as 76+2 millionths of a second per year. Since then, the observations have been improved further and show an agreement with Einstein's theory that has an error less than 1 per cent. This was a spectacular and comprehensive test of the General Theory, and effectively ruled out any other theory as a good description of the way the Universe works.

But astronomers were not prepared to rest on their laurels, and kept searching for other objects which might be used to test the General Theory. Their latest success involves a neutron star and a white dwarf star orbiting around each other some 7,000 light years from Earth. The neutron star – another pulsar, dubbed PSR J0348+0432 – was discovered by radio astronomers using the Green Bank Telescope, and its companion was soon detected in optical light, with the system being studied using both optical and radio telescopes around the world from late 2011. The two stars orbit around each other once every 2.46 hours, with the pulsar spinning on its axis once every 39 milliseconds – that is, roughly 25 times per second. The same kind of analysis as that used for the binary pulsar reveals that in this case the neutron star has a mass just over twice that of the Sun, with a diameter of about twenty kilometres, while the white dwarf has a mass a bit less than 20 per cent of the mass of the Sun. The distance between the two stars is about 1.2 times the radius of the Sun, just over half the Sun's diameter; so once again the whole system would fit inside the Sun. With the measured orbital properties, this implies that gravitational radiation should make the orbit 'decay' at a rate of 2.6×10^{-13} seconds per second; the measured rate is 2.7×10^{-13} seconds per second, with an uncertainty of $+ 0.5$.* Over a whole year, this amounts to just 8 millionths of a second. This is an even better test of the General Theory, partly because of the larger mass of the pulsar (the most massive neutron star yet discovered) compared with the neutron stars in the original binary pulsar system.

Over the years ahead, continuing observations will provide even more precise tests of the General Theory. But the accuracy of the test is already so precise, and the agreement with the predictions of Einstein's theory is so good, that the General Theory of Relativity can now be regarded as one of the two most securely founded theories in the whole of science, alongside quantum electrodynamics. So what can this spectacularly successful theory tell us about the Universe at large?

The Universe at large

The name 'General Theory of Relativity' actually has a double meaning. It is general because it applies to accelerated motion and gravity, not just to objects moving in straight lines at constant speed. This is the sense in which Einstein originally used the term. But it is also general in

---

* Don't worry about the bizarre units. Astronomers can't actually measure the change in one second, but derive this from observations made over much longer intervals. What matters is the close agreement between the two numbers.

the sense that it applies to everything – the entire Universe, all of space and time and all it contains. Indeed, strictly speaking the General Theory *only* applies to an entire universe. When we use it to describe the orbit of Mercury or the behaviour of a neutron star, what we are really doing is using an approximation in which the local metric is stitched onto the metric of surrounding flat spacetime, assumed to extend off forever. The influences of other parts of the Universe are ignored, because they are so small. But Einstein, as I describe in the next chapter when I get back to his life story, realised at once that what he had discovered was a description of the whole Universe, and in 1917 he published solutions of his equation, the first of the 'cosmological models', as these descriptions of spacetime are called, derived from the General Theory. Note the plural. There is no unique cosmological solution to the equations, but several possibilities, describing different kinds of universe. The lower case (universe) is used when talking about these models; the capital (Universe) is reserved for when we are discussing the Universe we live in.

This came as a surprise to Einstein. In 1917, he was looking for a unique solution to his equations, a description of *the* Universe. At that time, what we now know to be our home galaxy, the Milky Way, was thought to be the entire Universe, and the Milky Way seemed to be an essentially static collection of stars, eternal and unchanging on the largest scale, apart from random motions. The solution

Einstein found nearly matched this expectation, with one irritating defect. His model would not stay still. Depending on how you looked at it, it would either expand forever or contract. He solved the problem by adding another term to the equations, what became known as the 'cosmological constant', which had the sole purpose of holding the model still, to match the way he thought the Universe was. According to the physicist George Gamow, Einstein later described this as the 'biggest blunder' of his career, for reasons that will become clear.*

Hot on the heels of Einstein's discovery of what he thought was 'the' cosmological solution to the equations of the General Theory, the Dutch astronomer Willem de Sitter found another solution – another cosmological model. This was not a realistic description of our Universe because it contained no matter – it described mathematically empty, stationary spacetime. But when mathematicians did the equivalent of sprinkling bits of matter in to de Sitter's universe to see what would happen, they found a curious thing. It expanded, with space stretching to increase the distances between the particles of matter. And if you imagined a being on one of those 'test particles' monitoring light from the other test particles, the equations said that the light would be stretched

---

* *My World Line*, Viking, New York, 1970. Gamow was a good storyteller, although he sometimes exaggerated for effect. But this is the original (and only) source for one of Einstein's most widely-quoted remarks.

by the expansion, shifting it from the blue end of the spectrum to the red – a redshift. Nobody saw this as anything more than a mathematical curiosity in 1917.

The next big step forward came from a Russian mathematician, Alexander Friedman, in 1922. It was Friedman who was the first to realise that it was futile to seek a unique cosmological solution to Einstein's equations, and that he was dealing with a family of solutions, a whole variety of cosmological models. And instead of forcing those models to keep still by adding a cosmological constant, he let them do their own thing, and expand or contract if they wanted to. But for completeness he also included models with different cosmological constants.

Three kinds of model discovered by Friedman are particularly important. In one variation on the theme, the model universes expand forever, slowing down because of the gravitational influence of all the matter they contain, but never stopping. These are called 'open' models. Another set of models start out expanding, but slow to a halt and then fall back in on themselves. These are called 'closed' models. And in between these two varieties there is a special kind of universe which expands ever more slowly and gradually comes almost to a halt, but never re-collapses. These are called 'flat' models, because the overall spacetime of such a universe is flat, except for the dents caused by objects like stars.

Friedman's work did not attract much attention at the time, partly because he was Russian and cut off from other scientists in those troubled times, and partly because he died in 1925 and was not around to promote it. But in 1927, unaware of Friedman's work, Georges Lemaître, a Belgian astronomer (who also happened to be an ordained Catholic priest), published a similar analysis of the equations. By then, it had just been established that the Milky Way is just one galaxy in a vast sea of similar objects, and there had been a few measurements of redshifts in the light from some of these other galaxies. Lemaître suggested that this was evidence that the Universe is expanding, and developed the idea that it must therefore have been smaller in the past, with galaxies and stars squeezed together in what he called the 'cosmic egg'.

None of this attracted much attention until the American astronomer Edwin Hubble (the man who had proved that there are galaxies beyond the Milky Way), building on the work of Vesto Slipher and assisted by the superb observations made by his colleague Milton Humason, measured the redshifts and distances of many more galaxies at the end of the 1920s and into the 1930s, and came up with Hubble's law – that the redshift of a galaxy is proportional to its distance. This law works whatever galaxy you are viewing from, and does not mean that we are at the centre of the Universe. Indeed, it means that there is no centre to the Universe! With Lemaître's cosmic egg idea and the work of many astronomers and physicists (including Gamow) over the next four decades, this Big Bang model became established as a good description of our Universe, matching the Friedman/Lemaître expanding flat model with no cosmological constant. If Einstein had believed what the equations were telling him, he could have predicted this in 1917.

Up until the late 1990s, improved observations using better telescopes (both optical and radio) and instruments in space seemed to match the simplest 'Einstein universe' more and more accurately – the discovery of the cosmic microwave background radiation, a weak hiss of radio noise coming from all directions in space and interpreted as leftover radiation from the Big Bang itself, was particularly telling. The picture that was established was of a Universe that burst out from a superdense state (perhaps a singularity) in an initial phase of rapid expansion called inflation, followed by a gradually slowing expansion matching the flat Friedman models. One intriguing feature is that in order to make spacetime flat, and to account for the way galaxies move within clusters of galaxies, there has to be a lot more matter than we can see in stars and galaxies. To match the understanding of Big Bang physics, this matter cannot just be cold everyday matter, but has to be a kind of stuff not seen on Earth or in the stars. It was dubbed 'dark matter', and the search for dark matter is ongoing. Then, at the end of the 20th century, an even more exciting discovery was made.

With the best telescopes ever available, it was possible to measure redshifts and distances for galaxies very far away across the Universe. It turned out that these galaxies are receding from us, because of the stretching of space, slightly faster than they 'ought' to be, according to the simplest version of Hubble's law.* The explanation wasn't long in coming. Everything fits together if there is, after all, a small cosmological constant, a kind of springiness of space, affecting the expansion of the Universe. When the Universe was younger and more compact, gravity dominated the expansion, slowing it down in line with Hubble's law; but as the galaxies have got farther apart and matter has thinned out, the influence of gravity has weakened, while the cosmological constant has stayed the same, so that it is just beginning to overcome gravity and make the Universe expand faster. If this continues, the ultimate fate of the Universe is eternal expansion at an ever faster rate. Einstein's 'biggest blunder' turns out to be a key to understanding the Universe.

There's more. The cosmological constant is a form of energy, sometimes called 'dark energy'. The springiness of space contains energy, just as an ordinary compressed spring contains energy. And as Einstein taught us, energy is equivalent to mass. So this dark energy contributes to making spacetime flat. In fact, it is the *dominant* form of mass in the Universe. Using data from a satellite known as Planck, in honour of the founding father of quantum physics, in 2013 astronomers were able to announce a

---

* In some ways, it is better to think of this as us receding from distant galaxies faster than we are receding from nearby galaxies.

superbly detailed breakdown of the material content of the Universe. Just 4.9 per cent of the mass of the Universe (less than one twentieth) is in the form of what we think of as ordinary matter, the stuff we are made of and the stuff stars and planets are made of. Exactly 26.8 per cent of the mass of the Universe, more than five times the amount of everyday matter, is in the form of the still-mysterious dark matter. More than two-thirds, 68.3 per cent, of the Universe is in the form of the cosmological constant, aka dark energy. And the Universe is 13.8 billion years old. This astonishingly precise description of the entire Universe rests securely on the foundation of the General Theory of Relativity – a legacy which Einstein could never have dreamed of when he was grappling with the cosmological equations under conditions of extreme privation in Berlin in the second half of the First World War.

5 The Icon of Science

Personal problems; Fame; A last quantum hurrah;
Exile; Spooky action at a distance; The final years

To pick up the threads of Einstein's personal story, remember that he was alone in Berlin, starting a new life, when the First World War broke out. He had complete freedom to work as he wished, but nobody, at first, to look after the domestic side of things. This became increasingly important as the Allied blockade began its attempt to starve Germany into submission. It was in these circumstances, which became increasingly difficult as the war progressed, that Einstein completed his masterwork. And he only survived the war thanks to the support he soon received from his cousin Elsa.

Einstein, as I have mentioned, worked obsessively, slept only when he was exhausted, forgot to eat and neglected personal hygiene. He completed his General Theory in 1915 and almost immediately moved on to the cosmological implications. As if that were not enough, in 1916 he made another major contribution to quantum physics, using the methods of statistical mechanics (harking back to his early interest) to explain the behaviour of electrons 'jumping' between energy levels in atoms. This behaviour produces the lines in spectra which are the characteristic fingerprints of different elements (used, among many other things, to measure redshifts of galaxies in the expanding Universe). In explaining the nature of these jumps, Einstein derived the equation for black-body radiation – Planck's formula – in a new way, and also laid the foundations for an investigation of the way atoms could be stimulated into the emission of radiation. Einstein had returned to the puzzle of how light interacts with matter armed now with the model of the atom as a tiny central nucleus surrounded by a cloud of electrons, developed by the Dane Niels Bohr from the experimental work of the New Zealander Ernest Rutherford. Using the idea that electrons 'jump' from one energy level to another inside the atom as they emit or absorb light quanta (photons), he discovered how suitably energised atoms could, in principle, be made to release a pulse of light quanta all with the same wavelength, at the same time, as an energetic beam of pure light.

This became, decades later, the foundation of laser physics; the acronym 'laser' stands for Light Amplification by Stimulated Emission of Radiation. All this work was completed in 1916, but published in the *Physikalische Zeitschrift* in 1917, in a paper which also introduced the idea that photons – particles of light – carry momentum, just like everyday objects such as cricket balls. The effort, coming on top of the race to complete the General Theory before he was beaten to it, and combined with his self-neglect, nearly killed him. But all the while, Elsa was there, providing increasingly important support as the conditions in wartime Berlin deteriorated.

Personal problems

One reason why Einstein flung himself back into his work in the summer of 1916 was the continuing deterioration of his relationship with Mileva. He had visited Switzerland at Easter, chiefly to see his sons, but refused to have any contact with his estranged wife. This understandably angered Hans Albert, then twelve years old, and their father parted from the boys on bad terms. Mileva herself became ill, undergoing some sort of breakdown, which Einstein dismissed as either faked or psychosomatic, and still refused to see her. He wrote to his friend Besso, who tried to act as a go-between, that: 'She leads a worry-free life, has her two precious boys with her, lives in a fabulous neighbourhood, does what she likes with her time, and innocently stands by as the guiltless party.' (Which, to be fair, she was.) His reaction to outside problems was always to immerse himself in his work, which he did to such effect that he soon became seriously ill.

Everything we need to know about Einstein's state of health in 1917 has been summed up by a friend of his, Janos Plesch, who was also a physician:

> As his mind knows no limits, so his body follows no set rules ... he sleeps until he is wakened; he stays

awake until he is told to go to bed; he will go hungry
until he is given something to eat; and he eats until
he is stopped.[1]

The person who did the telling, of course, was Elsa,
assisted by her two grown-up daughters (Ilse was twenty
in 1917; Margot was eighteen).

At the beginning of 1917 Einstein suffered a physical
and mental collapse. He was only 38 in March that year,
but experienced severe stomach pains, which he initially
suspected were caused by cancer but which were diag-
nosed as due to an ulcer. The problem would affect him
severely for the next four years, and to a lesser extent for
the rest of his life. Over two months, early in 1917, he
lost 50 pounds in weight. A slow recovery only really
began in the summer of that year, when an apartment in
the building where Elsa and her daughters lived became
vacant and she was able to move Albert in there to take
full-time care of him. He might well not have survived the
worst months of 1917, when the Allied blockade was at
its most effective and food was tightly rationed, without
this support. The relationship was exactly what both of
them needed – Einstein needed someone to look after
him; Elsa needed someone to look after. As his health
improved, in 1918 the question of a divorce emerged once
again. Einstein made a generous offer to Mileva. He would
pay her 9,000 marks a year, 2,000 of which would be ear-
marked to go into a fund for the children, and 'The Nobel

Prize – in the event of the divorce and the event that it is bestowed upon me – would be ceded to you'. In 1918, the value of the Prize was 135,000 Swedish kronor, or 225,000 German marks. Even better, the kronor was a stable currency, whereas the mark was already showing signs of collapse. Negotiations over details took months, but the offer was eventually accepted and the divorce became final in February 1919 (ironically, on Valentine's Day), and he married Elsa in June.

It was also in 1919 that Einstein became world-famous outside the scientific community, when his light-bending prediction was confirmed by observations of a total eclipse of the Sun. But the public icon that he became looked very different from the dashing, dark-haired young man who had set out on the road to the General Theory in 1905. Illness and wartime privation had turned him into the white-haired professor that became 'the' image of a scientist (including Emmett Brown in the *Back to the Future* movies) until Stephen Hawking came along.

Fame

Among other things, as we saw in Chapter 3, the General Theory of Relativity predicts exactly how much space-time will be curved near a massive object like the Sun, and how light rays (which we are used to thinking of as travelling in straight lines) follow curved paths near the Sun as a result. This bending of light rays would show up from Earth, if we could look past the Sun at stars beyond the Sun, as a tiny sideways shift in the apparent positions of those stars on the sky – if the light from those stars wasn't overwhelmed by the brightness of the Sun. A total eclipse would occur on 29 May 1919, but there was little prospect of the necessary observations being made while the war raged. The fighting in Europe stopped just in time. In 1919, after the Armistice but before the formal peace treaty ending the war had been signed, a British expedition led by Arthur Eddington was sent to observe a solar eclipse visible from an island off the west coast of Africa; it confirmed the prediction of the General Theory, at least to Eddington's satisfaction, although in truth the observations were barely adequate. The fact that a German prediction had been confirmed by a British expedition so soon after the cessation of hostilities helped to ensure maximum publicity for the discovery, announced to a joint meeting of the Royal Society and the Royal Astronomical Society in London on 6 November 1919. The size of the 'shift' in the positions of the stars in the sky was, as predicted by Einstein, equivalent to the thickness of a matchstick seen at a distance of a little more than half a mile.

The news, as the headline writers put it, that Isaac Newton's theory of gravity had been superseded, that space was curved and that light could be bent as it passed by the Sun, made waves around the world and made Einstein famous. *The Times*, usually more sombre, ran the story under the headline 'Revolution in Science – New

Theory of the Universe – Newtonian Ideas Overthrown'.
The New York Times was no less enthusiastic:

> LIGHTS ALL ASKEW IN THE HEAVENS
> Men of Science More or Less
> Agog Over Results of Eclipse
> Observations
> EINSTEIN THEORY TRIUMPHS

Of course, Newtonian theory had not been 'overturned'. Newton's theory of gravity still applies in regions of weak gravity, and is the theory that space scientists use, for example, when planning the trajectories of spacecraft visiting the planets of the Solar System. You could do this using the field equations of the General Theory, but you would get exactly the same answer in the 'weak field approximation'. Einstein's theory goes beyond Newton's theory, but contains Newton's theory within itself. This is something that amateur theorists who delight in trying to find a better theory than Einstein's seldom appreciate (at least, judging by the mail I receive). Any theory of gravity, space and time that is better than the General Theory will have to include the General Theory within itself, explaining everything that the General Theory explains, including the behaviour of weightless gyros and binary pulsars, and then something more besides.

One result of Einstein's new-found fame was that his scientific work began to take a back seat. He had become the iconic figure that, until the Stephen Hawking phenomenon, provided the public with their image of what a great scientist 'ought' to be. But he wasn't finished with science yet. Einstein would make just one more important contribution to the quantum theory, in 1924 when he was 45 years old, and in the 1930s he would initiate a line of thinking which has profound implications today. But between 1919 and 1924, although he continued to carry out what was by his standards routine physics, he enjoyed some of the trappings of his fame, travelled and (especially after the award of the Nobel Prize) began to revel in the role of a kind of father figure to the rising generation of physicists.

Although economic conditions in Berlin were terrible after the war and there was political chaos, the presence of colleagues such as Planck to discuss physics with encouraged Einstein to keep his base there, even though he was offered a special professorship back in Zurich, in the comfortable safe haven of Switzerland. But that did not stop him travelling abroad at almost every opportunity when invited – he even agreed to visit Zurich to give a regular series of lectures each year, not least since payment in the stable Swiss currency would go a long way to ease his living conditions in Berlin. In 1921 he made his first visit to the United Sates, a lecture tour as part of a fund-raising programme for the Hebrew University in Jerusalem. His reception by the public resembled that of a modern pop star, and his lectures on relativity theory drew packed houses, while he was also awarded civic receptions and honorary degrees.

Considering his global status by 1921, it might seem that the Nobel Committee were rather tardy in awarding Einstein their Physics Prize. In fact, he had been nominated every year since 1910 except for 1911 and 1915. Most of the people who did receive the award in those years thoroughly deserved it, although the curious exception is Niles Dalén, a Swedish inventor given the 1912 Prize for his automatic regulator to control gas-fuelled lighthouse lamps. The same year, Einstein was nominated unsuccessfully for the Special Theory of Relativity. And in 1916, when Einstein was nominated for his work in molecular physics, no award was made. In 1921, the award was deferred, and then awarded to Einstein the following year, while Niels Bohr received the 1922 Prize. The citation for Einstein's award referred to 'services to theoretical physics, and especially for his discovery of the law of the photoelectric effect'. But the award was presented 'without taking into account the value that will be accorded your relativity and gravitation theories after these are confirmed in the future'. One way of interpreting this is that the Nobel Committee still weren't convinced of the importance of the General Theory; a more generous interpretation, which I favour, is that they were carefully leaving the door open for the possible award of a second Prize once more tests had fully confirmed the validity of his theory of gravitation. Either way, no such second award was ever made, not least because over the next two decades there would be a wealth of deserving physicists receiving Nobel Prizes for their work on quantum physics. So, although Einstein did receive the Nobel Prize, it was specifically not for his masterwork! Not that this mattered to Mileva and the boys, who received the financial rewards while Albert had the glory.

At the time the award was announced, the Einsteins were in Japan. He was there on a lecture tour as a guest of the publishing house Kaizosha, following a successful tour by Bertrand Russell the previous year. Before Russell left Japan, his hosts asked him to name the three most important living people, so that they could be invited to follow in his footsteps; Russell gave them just two names, Lenin and Einstein. Since Lenin was busy at the time, they invited Einstein. The invitation was particularly welcome as it meant Einstein would be away from Germany for at least six months, from October 1922, at a time when the political situation was deteriorating following the assassination by right-wing extremists of Walther Rathenau, a Jew who was the Minister of Reconstruction. This was part of a pattern of growing anti-Semitism and violence.

By the time the Einsteins returned to Europe in 1923, his fame had reached even greater heights, thanks to the award of the Nobel Prize, but Germany was plunging deeper into the depths of economic collapse, runaway inflation and violence, which paved the way for the rise of Nazism. Amidst this turmoil it is doubly surprising that at the relatively grand old age (for a theoretical physicist) of 45 Einstein would still produce one last piece of really significant science. But he did not do it alone.

A last quantum hurrah

Einstein's fame and status in the scientific community meant that he was sent streams of communications from aspiring scientists, as well as from his established colleagues, from all over the world. In June 1924 he received a letter from a young Indian physicist, Satyendra Bose, based in the city then known as Dacca.* The letter accompanied an unpublished scientific paper in which Bose developed Einstein's ideas on light and radiation in new ways. Bose had found a new way to derive the equation for black-body radiation, without assuming that light behaved as a wave at all but simply as a quantum gas obeying a new kind of statistical law. Einstein had cut his scientific teeth on statistical laws, and immediately saw the importance of this discovery. He first translated Bose's paper (which was written in English) into German and got it published in the *Zeitschrift für Physik*, then picked up the idea and developed it further himself, applying the new statistics not just to particles of light but, under appropriate circumstances, to molecules and atoms.

---

* This was before the partition of India and the creation of Pakistan; Dhaka is now the capital of Bangladesh.

The new statistical technique soon became known as 'Bose-Einstein statistics', and particles that obey Bose-Einstein statistics are called 'bosons'. Einstein was able to use the new statistics to make predictions about the nature of thermodynamics, and in particular the behaviour of certain liquids at very low temperatures, where viscosity disappears and they become 'superfluid'; the predicted superfluidity was observed experimentally in 1928, and the behaviour of bosons, or 'bose condensates', is still a matter of extreme interest to researchers today.

But the most far-reaching aspect of Einstein's work in the mid-1920s came from the analogy between a quantum gas and a molecular gas. Einstein realised that it worked both ways. If the black-body radiation behaved in the same way as a molecular gas under some circumstances, and if light quanta were also known to have a wave-like nature, the implication was that molecules and other material objects must also have a wave-like nature.

Einstein published this conclusion, derived from Bose-Einstein statistics, in 1925; but there was no need for him to take it further and develop a complete wave theory of matter because someone else had already come up with the same idea from a different angle. The Frenchman Louis de Broglie had finished his PhD thesis in Paris in the spring of 1924 and submitted it to Paul Langevin, an old friend of Einstein. De Broglie was older than the average PhD student, having been born, at Dieppe, on 15 August 1892. He came from an aristocratic family and initially studied history at the Sorbonne, entering in 1909, being intended for a career in the diplomatic corps. But under the influence of his elder brother Maurice (seventeen years his senior), who became (very much against the wishes of his father) a pioneering researcher interested in X-ray spectroscopy, Louis began studying physics alongside history. Maurice had obtained his doctorate in 1908, and had been one of the scientific secretaries at the first Solvay Congress. Louis' study of physics was interrupted in 1913 by what should have been a short spell of compulsory military service, but was extended when war broke out. During the First World War, he served in the radio communications branch of the army, operating for a time from the Eiffel Tower. It was because of army service and his switch from history that de Broglie's scientific studies were delayed, so that his PhD thesis was only submitted (at the Sorbonne) in 1924, when he was in his thirties. But by then he had already published several important papers on the properties of electrons, atoms and X-rays, and he was one of the first physicists to accept fully the idea of light quanta, which he discussed in an article published in 1922.

The thesis developed from Albert Einstein's earlier work – the work for which he had received the Nobel Prize, which showed that light (traditionally thought of as a wave) could also be explained in terms of particles (now known as photons) – to propose that 'particles' could also behave as waves, with the wave and particle aspects of the quantum entity linked by the equation wavelength × momentum = h, where h is Planck's constant. De Broglie's supervisor, Paul Langevin, was nonplussed by this and showed the thesis to Einstein. Einstein (who had just received Satyendra Bose's paper on light quanta) wrote back to assure Langevin that de Broglie's work was sound: 'I believe,' he said, 'that it involves more than a mere analogy.' De Broglie duly received his PhD. In his second paper on the 'Bose gas', Einstein made a reference to de Broglie's work which caught the attention of Erwin Schrödinger and started him down the road that led to wave mechanics.[2]

In plain English, in his thesis, de Broglie suggested that all material particles (electrons and the like) had a wavelength. Five years later, de Broglie received the Nobel Prize for his work.* By then, largely thanks to this breakthrough realisation that matter particles also have a wave-like nature, quantum theory was fully established, and Einstein had been in at both the beginning and end of the story. But in 1928 he had suffered another serious illness. In 1929, the year de Broglie received his Nobel Prize, Einstein was 50 years old, no longer a major player in the science game, and beginning to realise that the situation in Germany might soon become untenable.

---

* Not least because the wavelengths of electrons had been measured by George Thomson in 1927.

Exile

Although Einstein continued to carry out research and write scientific papers after his 50th birthday, hardly surprisingly he did nothing else to rank with his earlier achievements. Most of his later scientific life was devoted to an unsuccessful effort to find a single mathematical package (a unified theory) that would describe both the material world and the world of electromagnetic radiation, echoing the way Maxwell had unified the description of electricity and magnetism into one package. It was a noble effort, but doomed to failure given the limited understanding physicists had of the particle world, in particular, at the time. We shall not describe any of this work, but sketch the outlines of Einstein's later years and mention just one influential idea that he had – although it did not work out as he had expected.

Clinging to the hope that the political situation in Germany might yet improve, Einstein avoided resigning from his post in Berlin as long as he could, but spent as much time as possible out of the country over the next few years. After spending the summer of 1930 in Berlin, he visited Brussels, London and Zurich (where he received an honorary doctorate from the ETH) in the autumn, returning briefly to Berlin before sailing for California from Antwerp on 2 December. The long voyage via New York and the Panama Canal ended in San Diego on 30 December, and Einstein stayed as a visiting professor at Caltech, in Pasadena, until mid-March 1931.

Once again, the return to Berlin was brief. In May, Einstein took up an invitation to spend a month giving a series of lectures in Oxford. The visit proved so successful that it led to an invitation for him to become a visiting fellow at one of the Oxford colleges, Christ Church, for five years, with an annual stipend of £400. He could come and go as he pleased, provided he spent some time each year in Oxford. As everyone involved realised, apart from the intrinsic attraction of such an offer, it provided an escape route if things got worse in Germany.

At the same time, Einstein was being courted by Caltech, with an offer of $7,000 for another visit the following winter, and an understanding that the arrangement could be made permanent if he wished. Once again, the Einsteins sailed on 2 December, and once again they arrived in California at the end of the year. It was during this visit that Einstein made the contact that would soon result in him finding a settled base.

The contact was with Abraham Flexner, a highly regarded American scientific administrator, who had obtained funding for a new research institute in Princeton and was on a headhunting expedition to the West Coast looking for eminent scientists to staff it. Who better than the most eminent scientist of the 20th century? Flexner realised that if he could lure Einstein to his new institute, his presence would act as a magnet for other scientists and ensure the success of the project. But he adopted a softly, softly approach to snaring his prize.

Einstein returned to Europe in March 1932, committed to spending the next winter at Caltech but with no plans for a permanent move. In the summer, while he was in Oxford, he met up with Flexner (still headhunting!) again. This time, the American offered Einstein a post at what would become the Institute for Advanced Study, on whatever terms Einstein wished. In June, a deal was agreed. Einstein would visit Princeton for six months each year, starting in the autumn of 1933, for a salary of $10,000 and all travelling expenses to be paid by the Institute. The arrangement came just in time. Following elections in July 1932 the Nazis became the largest party in the German government, and the writing was on the wall. Officially, when Albert and Elsa left for their third and last visit to Caltech in December 1932 Einstein was expected back in Berlin to take up his academic duties in April 1933, but as he locked the door of their home, Einstein turned to his wife and said: 'Take a very good look at it. You will never see it again.'[3]

Adolf Hitler was appointed Reich Chancellor on 30 January 1933, and in the ensuing wave of anti-Semitism Einstein's bank account was frozen, his house was ransacked and copies of a popular book he had written on relativity were publicly burned. Although the Einsteins returned to Europe at the end of March, they clearly could not return to Germany, but largely divided their time between Oxford and Belgium before leaving for Princeton in October. This was still not seen, at the time, as a permanent move, and it was expected that Einstein would return to Oxford the following summer. There were even moves to grant him British citizenship. In fact, after 1933 Einstein only left the United States once, travelling to Bermuda in May 1935 in order to make a formal application to re-enter the country as a permanent resident.

Soon after taking this step, the Einsteins were able to buy a house in Princeton, 112 Mercer Street – they had previously been living in a rented apartment. But Elsa did not enjoy the new property for long. She became ill during the summer and never fully recovered. She died in the house on Mercer Street on 20 December 1936. Einstein, always independently-minded, seems to have quickly got over the loss. By this time his divorced step-daughter Margot was acting as his housekeeper and looking after him, while he had an efficient secretary, Helen Dukas, who protected him from outside intrusions. As he wrote to his friend Max Born not long after Elsa died:

> I have settled down splendidly here. I hibernate like
> a bear in its cave, and really feel more at home than
> ever before in all my varied existence. This bearish-
> ness has been accentuated further by the death of
> my mate, who was more attached to human beings
> than I.[4]

Spooky action at a distance

The year before Elsa died, Einstein made his last important contribution to science, although he would have been astonished at the way an idea he presented in 1935 was developed, and at its importance in the applied physics of the 21st century.

The idea, known from the initials of the authors of the paper in which it was presented as 'the EPR experiment' (although Einstein was the brains behind it) was originally put forward in 1935 by Einstein and two of his colleagues in Princeton, Boris Podolsky and Nathan Rosen. It was a 'thought experiment' (like the freely falling lift) intended to demonstrate (as they thought) the logical impossibility of quantum mechanics. Einstein had no expectation that such an experiment could, or would, ever be carried out. But the basic idea was adapted by David Bohm in the 1950s and refined by John Bell in the 1960s to become a practicable experiment that was actually carried out in the 1980s, most notably by Alain Aspect and his team in Paris, establishing that nature really does behave in a non-commonsensical way.

In the original version of the thought experiment – sometimes referred to as the 'EPR paradox', although it is not really a paradox – the EPR team imagined a pair of particles that interact with one another and then separate, flying far apart and not interacting with anything else at all until the experimenter decides to look at one of them. At the time the particles interact, it is possible to measure the total momentum of the system, and this cannot change if they do not interact with anything else. So if the experimenter chooses, much later, to measure the momentum of one particle, it is possible to calculate the momentum of the other particle, far away, by subtracting the measured momentum from the total. We know that quantum physics requires that by measuring the momentum of the first particle we destroy information about its position, because of Heisenberg's uncertainty principle. But the EPR team suggested that quantum uncertainty could be circumvented by measuring the momentum of the first particle and the position of the second particle, while calculating the momentum of the second particle in the way we have outlined. The only alternative would be that by measuring the momentum of the first particle we destroy information about the position of the second particle (or prevent such information ever existing), instantaneously, no matter how far apart they are.

Einstein referred to this as a 'spooky action at a distance', arguing that it was both logically absurd and impossible for any communication to travel faster than light, so quantum theory must be flawed. But the experiments show that this kind of 'non locality' is indeed a feature of the quantum world, and that measurements made on one particle of such a quantum pair really do affect its counterpart, instantaneously, no matter how far away that counterpart may be. This is the opposite of what Einstein (and, indeed, John Bell) expected. Even more profoundly, the experiments show that non-locality is not solely a feature of quantum theory, but a feature of the physical world. Whatever theory is the best description of reality, it must include non-locality, just as any satisfactory theory of gravity must include the inverse square law. And this is more than abstract theorising. This 'entanglement' of quantum entities is now being put to practical use in the infant technology of quantum computing, in developing uncrackable quantum codes, and (soon) a totally secure quantum internet.[5] It has become one of the most enduring and important parts of Einstein's legacy, albeit in the opposite way from what he had in mind. But it really was his scientific swansong.

The final years

Over the next few years, most of what was left of Einstein's family also came to America. Hans Albert, who had completed a PhD in engineering at the ETH in 1936, arrived in 1937 with his wife and son and settled in the United States; he died in 1973. But Eduard, who had developed signs of serious mental illness in the early 1930s, ended his days (in 1965) in a psychiatric hospital in Switzerland. In 1939, Maja and her husband Paul Winteler had to leave Italy, where the Fascists were in power. Paul was refused entry to the United States and stayed in Geneva; Maja joined the household in Princeton.

Also in 1939, Einstein played his famous part in alerting the American President, Franklin D. Roosevelt, to the prospect of an atomic bomb. The historic letter to Roosevelt was actually drafted by other scientists, concerned that Hitler's Germany might develop atomic weapons, but Einstein was persuaded to sign the letter and send it to the President since his name would carry more weight. Very little happened as a result of the letter before the Japanese attack on Pearl Harbor in 1941, but this was the first step in the United States towards the development of the atomic bomb, although Einstein (who became a US citizen on 1 October 1940) played no part in the Manhattan Project itself. This was partly because the authorities involved had doubts about his discretion and were concerned about the left-wing and pacifist sympathies he had expressed earlier in his life. But Einstein had long been aware of the possibility of using nuclear energy as a source of power to replace coal and oil. In 1920, he discussed Ernest Rutherford's then recent work on 'splitting the atom' with Alexander Moszkowski, and commented: 'It seems feasible that, under certain conditions, Nature would automatically continue the disruption of the atoms, after a human being had intentionally started it, as in the analogous case of a conflagration which extends, although it may have started from a mere spark.' This is what became known as a 'chain reaction'.

Einstein's contribution to the American war effort was limited to acting as a consultant for the US Navy, assessing various schemes put forward for new weapons. This was an ideal job for a former Technical Expert in the Swiss patent office, but perhaps did not make full use of his abilities.

After the war, Einstein experienced another bout of serious illness and was never fully fit. He officially retired in 1945, but kept his office at the Institute and continued to work there whenever he wanted to and felt up to it. Maja had intended to go back to Switzerland when the war ended, but she suffered a stroke and became bedridden; Einstein read to her every day until her death in 1951. By then, Mileva had already died, in Zurich in 1948. It's hardly surprising that when the famous offer of the presidency of Israel came in November 1952, Einstein, now 73, felt unable to accept. In his formal letter turning down the invitation he said: 'I lack both the natural aptitude and the experience to deal properly with people and exercise official functions ... even if advancing age was not making increasing demands on my strength.'[6]

In spite of his physical deterioration, however, he remained mentally fit, and tried to use the power of his name to nip the nuclear arms race in the bud. After the announcement of the American hydrogen bomb programme in 1950, Einstein made a televised broadcast in which he warned that unless the continuing development of bigger and 'better' bombs were stopped:

Radioactive poisoning of the atmosphere and, hence, annihilation of all life on Earth will have been brought within the range of what is technically possible. The weird aspect of this development lies in its apparently inexorable character. Each step appears as the inevitable consequence of the one that went before. And at the end, looming ever clearer, lies general annihilation.[7]

Right to the end of his life, Einstein continued to speak out against the nuclear arms race and in defence of the civil liberties attacked in the early 1950s by the McCarthy witch hunts. But that end was not far off. He became ill again in April 1955, not long after his 76th birthday. A month after the 50th anniversary of the completion of the first paper of his *annus mirabilis* (the paper for which he received the Nobel Prize), and seven months short of the 40th anniversary of the presentation of his masterwork, Einstein was taken to hospital. He refused any treatment to prolong his life, describing such intervention as 'tasteless'.[8] A little after 1am on 18 April 1955, with only a nurse in attendance, he muttered few words of German and died. The nurse knew no German.

Further Reading

Most of these books are accessible at about the level of the present volume but go into more detail about Einstein's life or work. Titles marked with an asterisk require a little more scientific background. Quotes in the text, unless otherwise indicated, are from the collected works or the Princeton archive. See *The Collected Papers of Albert Einstein*, volumes 1–10, published by Princeton University Press between 1987 and 2006. These take the story up to 1920, covering the major part of the story told in this book.

Amir Aczel, *God's Equation*, New York: Random House, 1999.

Jeremy Bernstein, *Albert Einstein and the Frontiers of Physics*, Oxford University Press, 1996.

Alice Calaprice, editor, *The Ultimate Quotable Einstein*, Princeton University Press, 2010.

Ta-Pei Cheng, *Einstein's Physics*, Oxford University Press, 2013.

Ronald Clark, *Einstein: The Life and Times*, London: Hodder & Stoughton, 1973.

Albert Einstein, *Ideas and Opinions*, New York: Random House, 1954.

Albert Einstein, *Relativity*, New York: Crown 1961 (reprint in English of Einstein's only 'popular' book; originally published by Holt, New York, 1921).

Albert Einstein, *Autobiographical Notes*, edited and translated by P. A. Schilpp, La Salle, Illinois: Open Court, 1979.

*Albert Einstein, *The Collected Papers*, Princeton University Press, 1987–2006 (see especially *Volumes 1, 2* and 6, 1987, 1990 and 1997).

Lewis Carroll Epstein, *Relativity Visualized*, San Francisco: Insight Press, revised edition 1987.

Albrecht Fölsing, *Albert Einstein*, translated by Ewald Osers, New York: Viking, 1997.

George Gamow, *Mr Tompkins in Paperback*, Cambridge University Press, 1965.

John Gribbin, *In Search of Schrödinger's Cat*, New York: Bantam, 1984.

John Gribbin, *In Search of the Edge of Time*, London: Bantam, 1992.

Mary Gribbin and John Gribbin, *Time Travel for Beginners*, London: Hodder, 2008.

Tony Hey and Patrick Walters, *Einstein's Mirror*, Cambridge University Press, 1997.

Walter Isaacson, *Einstein*, New York: Simon & Schuster, 2007.

Thomas Levenson, *Einstein in Berlin*, New York: Bantam, 2003.

Robert Millikan, *The Autobiography*, London: Macdonald, 1951.

Alexander Moskowski, *Conversations with Einstein*, translated by Henry Brose, London: Sidgwick & Jackson, 1970 (reprint of 1921 edition).

Dennis Overbye, *Einstein in Love*, New York: Viking, 2000.

*Abraham Pais, *Subtle is the Lord*, Oxford University Press, 1982.

Jürgen Renn and Robert Schulmann, ed., *Albert Einstein, Mileva Maric: The Love Letters*, translated by Shawn Smith, Princeton University Press, 1992.

John Rigden, *Einstein 1905*, Cambridge University Press, 2005.

Carl Seelig, *Albert Einstein*, translated by Mervyn Savill, London: Staples Press, 1956.

John Stachel, ed., *Einstein's Miraculous Year*, Princeton University Press, 1998.

Russell Stannard, *The Time and Space of Uncle Albert*, London: Faber, 1989.

Michael White and John Gribbin, *Einstein: A Life in Science*, London: Simon & Schuster, revised edition 2005.

Clifford Will, *Was Einstein Right?*, New York: Basic Books, 1986.

Other biographies by John and Mary Gribbin

Richard Feynman: A Life in Science, London:
    Penguin, 1998.

*FitzRoy: The Remarkable Story of Darwin's Captain and the
    Invention of the Weather Forecast*, New Haven: Yale
    University Press, 2004.

*He Knew He Was Right: The Irrepressible Life of James
    Lovelock*, London: Penguin, 2009.

Endnotes

Chapter One: In the Beginning

1. *Collected Papers*.
2. Quoted by, for example, Dennis Overbye.
3. See Seelig.
4. See Overbye.
5. *Collected Papers*.
6. *Collected Papers*.
7. *Albert Einstein, Mileva Maric: The Love Letters*.
8. *Collected Papers*. See also Fölsing.
9. See Fölsing.
10. Article in the *New York Post*, 25 February 1931.
11. *Collected Papers*.
12. See Overbye.
13. *Collected Papers*.
14. See Overbye.

Chapter Two: The *Annus Mirabilis*

1. See Seelig.
2. See, for example, Fölsing.
3. *Collected Papers*; Kleiner translation from Fölsing, Burkhardt translation from Stachel.
4. Tony Cawkell and Eugene Garfield, 'Assessing Einstein's impact on today's science by citation analysis' in *Einstein: The First Hundred Years*, edited by Maurice Goldsmith, Alan Mackay and James Woudhuysen, Oxford: Pergamon, 1980.
5. In the *Philosophical Magazine*.

6. Quoted by John Heilbron in his biography of Planck: *The Dilemmas of an Upright Man*, Berkeley: University of California Press, 1986.

7. Interview quoted by Fölsing.

8. *Collected Papers*.

9. *Physikalishel Zeitschrift*, vol. 17, p. 217, 1916.

10. *Collected Papers*.

11. *Collected Papers*.

Chapter Three: The Long and Winding Road

1. Quoted by Fölsing.

2. Comment made to Max Born; see, for example, Fölsing.

3. See Pais.

4. See Clark.

5. See Isaacson.

6. See *Ideas and Opinions*.

7. *Collected Papers*.

8. See Isaacson.

9. See Calaprice.

Chapter Four: Legacy

1. *Physical Review*, vol. 56, pp. 455–9.

2. *Science*, 26 April 2013, vol. 340, no. 6131.

Chapter Five: The Icon of Science

1. See *Janos: The Story of a Doctor*, by John Plesch, London Books, 1947.

2. See my book *Erwin Schrödinger and the Quantum Revolution*, London: Bantam, 2012.

3. Quoted by Philipp Frank, in *Einstein: His Life and Times*, translated by George Rosen, New York: Knopf, 1947.

4. Albert Einstein and Max Born, *The Born-Einstein Letters*, London: Macmillan, 1971.

5. See my book *Computing with Quantum Cats*, London: Bantam, 2013.

6. Otto Nathan and Heinz Norden, ed., *Einstein on Peace*, New York: Simon & Schuster, 1960.

7. See *Einstein on Peace*.

8. See Pais.

Index

awarded to Marie and Pierre
Currie 43
awarded to Robert Millikan 91
awarded to Russell Hulse and
Joseph Taylor 170
non-Euclidean geometries 37–8,
138–40, 144–5
nuclear arms race 111, 207–8
nuclear energy 206

Olympia Academy 37, 40, 43
Oppenheimer, Robert 158
'Oppenheimer-Volkoff limit' 158
osmosis 62
osmotic pressure 52–3, 61–3
oxygen 18, 50, 53, 60

Pais, Abraham 127
particles
charged 75, 98
of light 68–93, 186, 195–8. *See
also* light
of matter 18–9, 41, 46, 49–51,
56–61, 64–7, 97–8, 108, 160,
178, 195–8, 203–4. *See also*
matter
vertical distribution of 67
Perrin, Jean Baptiste 67–8
photoelectric effect 81, 84, 87–9,
90–3
photons 47, 69, 85–6, 91–2, 186,
197. *See also* atom and
quanta
Physical Review Letters 168
Physics Today 94
physics
'classical' 123, 147

laser 186
molecular 193
quantum 30, 57, 68, 93, 122,
129, 135, 182, 185, 194, 204.
See also quantum
Physikalische Zeitschrift 186
Planck, Max 30, 43, 74–81, 85–92,
107, 110–13, 120–2, 126,
186, 196–8
Planck's constant 77, 85, 91,
198
Planck's equation 76–7, 81, 85,
88, 91, 107, 186, 198
'Planck' satellite 182
Plesch, Janos 187
Podolsky, Boris 203
Poincaré, Henri 37–9, 99–101, 104,
107, 111, 132, 139
Science and Hypothesis 37
postulates 101–6
Pound, Robert 163
Prague 126–33, 137
Princeton University 3, 160,
200–5
Prussian Academy of Sciences
133–4, 150–5
pulsars 159, 171–6
binary 170–2, 176, 191
orbital period 171–4
Pythagoras' Theorem 6, 100

quanta 30, 43, 80–2, 84–6, 88–9,
91–2, 110–14, 120, 130, 186,
196–8
light 43, 47, 82, 85, 89, 91–2,
110, 113, 12–3, 186,
196–8